THE
RESISTOR HANDBOOK

Second Edition

Cletus J. Kaiser

Published
by

CJ Publishing
398 Wintergrape Ln.
Rogersville, MO 65742

ISBN: 0-9628525-5-4
Library of Congress Catalog Card Number: 98-92626

Printed in the United States of America.

Table of Contents

Acknowledgements

The author is thankful to The Lord.

The author is deeply indebted to his family for their guidance and support.

Through the courtesy of Matthew Pobursky, the author gratefully acknowledges his professional help in making this book possible.

Appreciation is also express to the many technical friends who suggested content and organization of this book.

Preface

The response from "The Capacitor Handbook" convinced me that resistors are also a misunderstood and misused component. This book complements "The Capacitor Handbook" by providing practical guidance for the use of resistors in electronic and electrical circuit design.

All chapters are arranged with the theory of the construction discussed first, followed by circuit application information. With all chapters arranged in the same manner, this will make reading and using this book for reference easier.

The first chapter covers the fundamentals of resistors. Each following chapter addresses a different resistor type. This book could have been titled: 'Everything You Always Wanted To Know About Resistors, But Were Afraid To Ask...'

This second edition was published to expand the information on metal foil resistor technology. A new chapter has been added to fulfill this request.

Since the first edition was published, "The Inductor Handbook" has also been published to complete The Passive Trilogy.

Chapter 1

Fundamentals For All Resistors

Power Rating and Thermal Dissipation

Resistors are energy convertors converting electrical energy to thermal energy in accordance with the formula $P=EI$, where the power consumed P watts is the result of impressing E volts across the resistor and producing I (amps) current. Also from Ohm's law, these additional formulae are recalled here for convenience: $P=I^2R$, $P=E^2/R$, and one watt-hour is equivalent to 3.413 BTUs. The energy converted by the resistor cannot be allowed to continually build up and be stored as heat within the resistor. Eventually it must be passed to an additional "other medium" for disposal. Equilibrium is eventually reached such that the heat generation within the resistor is equal to the heat disposal to the other medium and a stable internal temperature results. This temperature must not exceed a tolerable limit since further increase will alter the integrity of the resistor or even destroy it. Since this temperature limit is within the resistor, it is referred to as the "maximum internal hot-spot temperature" as if there were one spot within the device that is most likely to overheat.

The internal hot-spot temperature must be limited to that which will not cause a permanent performance change through metallurgical degradation in a metal alloy resistive element or a chemical change in a cermet or other nonmetallic element. More often than not, it is limited to the lower temperatures that will be tolerated by any coatings in contact with the element.

From the internal hot-spot the heat flows through the body of the resistor and (in the absence of a heat sink) passes to the other medium by radiation and convection from the resistor body. With a heat sink, more rapid removal by conduction takes place. The leads act as thermal conductors to the board traces (which act as heat sinks) and dissipate a major quantity of the total heat from the resistor.

The equilibrium temperature will be reached most rapidly with the resistor in still air. Additional power can be dissipated before reaching the maximum temperature if there is circulation of the air. A heat sink or an oil bath or cooling water circulation becomes necessary to remain below the maximum internal hot-spot temperature when dissipating increasingly large amounts of power.

The resistor manufacturer has determined the maximum power in still air and/or with a specified heat sink that will limit the resistor internal hot-spot temperature to a satisfactory level. This is the rated power and must not be exceeded. A separate power rating for still air and for a specified heat sink may be provided. Also, the other medium may not always be at room temperature and therefore less power can be dissipated if the other medium is at an elevated temperature. A derating curve may be provided allowing use of the resistor in an elevated temperature environment without exceeding the maximum internal hot-spot temperature. These curves generally derate to zero power at some elevated temperature which is for these purposes "the maximum hot-spot temperature."

While manufacturer's power ratings are serious specifications, they cannot be expected to exactly apply in every case. The thermal dissipation in critical applications may require further analysis. To limit the amount of calculation required, the temperature drop from the "internal hot-spot" to some "test spot" on the body of the resistor is frequently known and available. The thermal analysis is thereby limited to factors associated with the other medium and confirmed by surface pyrometer readings on the "test spot." This "test spot" with its "maximum external hot-spot temperature" is frequently identified for simplicity as the "hot-spot."

One Short Pulse

The theory of pulse handling depends upon the pulse width. One short pulse of 100 milliseconds or less is assumed to never have time enough to do more than heat the element. Therefore the calculation is based on the total mass of the element (wire) only being heated to the maximum internal hot-spot temperature.

Example: A resistor element consisting of 10 grams of wire with a specific heat of 0.1 heated from 25°C to a maximum internal hot-spot temperature of 275°C requires:

Note: A calorie is 1/860 watt-hours.

$$\frac{10 \times 0.1}{860} (275-25) = 0.29 \text{ watt-hours of equivalent power}$$

If the pulse width is 10 milliseconds the pulse power could be:

$$\frac{0.29 \times 3600 \times 1000}{10} = 104,400 \text{ watts max.}$$

Equally Spaced Repetitive Pulses

With equally spaced repetitive pulses the average power which is the peak power multiplied by the pulse width divided by the cycle width establishes a threshold temperature rise. Above this threshold to the maximum internal hot-spot temperature is additional pulse capacity. In reducing this theory to practice, assignment is given to the percent of rated power that both the average and pulse power represent, and as long as the sum of the two do not exceed 100% of the rating, then the application is satisfied.

Example: The pulse power is $P=E^2/R$ and the average power is $P_A=Pt/T$ where t is the pulse width and T is the cycle width. Calculate the average power and the percent of rated power for the candidate resistor. Now using the derating curve establish the threshold temperature that results and use it in the pulse calculation above. Calculate the pulse power that can be tolerated in the wire being heated above the new threshold temperature to the internal hot-spot temperature. This is the maximum pulse power allowed at this average power level.

Long Pulses

With long pulses (100 milliseconds to 5 seconds) the pulse heat generated is no longer retained fully within the wire but some passes to other body members of the resistor. A short time overload rating is frequently given and this can be safely used as the long pulse rating. A factor of 5X rated power is generally satisfactory for 5 seconds.

Allowable pulse power levels for pulses from 1 to 25 seconds long can be found by 25X rated power divided by the pulse time in seconds. Below 1 second, use the 1 second rating. Above 25 seconds, use rated power.

Current Rating

Many resistive shunt and current sensor devices are "current limited" rather than power limited. This comes from the fact that as electron flow is increased by increasing the voltage, there comes a point where the current no longer follows Ohm's law. This "critical current density" varies for different resistance materials but it must be kept in mind that leads and attachment hardware also have a critical current density. A current density up to 260 KA/in^2 may be tolerated by a particular metal but the manufacturer's rating is likely to be considerably below this limit for safety reasons.

Temperature Coefficient of Resistance (TCR)

Change in resistance as a function of temperature exists in all resistors to various extents. This change is seldom linear but for convenience it is usually expressed as a straight line function. Sometimes the change is sought between two specific temperatures. Hence the general equation for the instantaneous TCR at any temperature is:

$$TCR_{inst} = \frac{dR}{Rdt}$$

The more useful TCR calculated from the chord slope from one temperature to another is:

$$TCR_{chord} = \frac{R_2 - R_1}{R_1(T_2 - T_1)}$$

Also, TCR is usually referenced to room temperature (24°C) as the reference temperature T1 and the second temperature T2 is either 0°C or +60°C for end use in instrumentation and -55°C or +125°C for military end use (Power resistors to +275°C). Note that TCR can be either positive or negative and by convention an increase in resistance with an increase in temperature is a positive TCR. Also note that self heating causes a resistance change due to TCR. The TCR unit of measure is ppm/°C (parts per million per degree Celsius). Figure 1.1 gives the TCR for some resistance materials.

Temp. Range	-55 to +25°C	0 to +25°C	+25 to +60°C	+25 to +125°C
Manganin	+50	+10	-5.0	-80
Zeranin	+20	±2.5	±5.0	+10
Evanohm	+5.0	+2.5	-2.5	-5.0
Foil	-1.0	-0.3	+0.3	+1.0
Thin Film	-10	-5.0	+5.0	+10
Thick Film	-100	-25	+50	+100

Figure 1.1 TCR, ppm/°C of Various Element Materials

Reactance/Impedance

Reactance - "That component of an AC impedance defining a phase shift between the current and the voltage."

Impedance - "The ratio of effective voltage over effective current in an AC circuit."

Reactance may be inductive or capacitive. In low value resistors (below 10 ohms) the inductive reactance usually outweighs the capacitive reactance while higher values are more likely to be capacitive with bifilar winding. Pi and bifilar winding are used to improve reactance.

Figure 1.2 shows the various components of resistor response to an AC signal in vectorial notation. "R" is the DC resistance of the device and "X_L" and "X_C" are the inductive and capacitive reactances respectively.

Figure 1.3 shows the sum of "X_L" and "X_C" leaving an effective net reactance "X_T." The impedance "Z" is the Pythagorean sum of "R" and "X_T." The AC impedance is always greater than the DC resistance unless "X_L" is exactly equal to "X_C."

Low value resistive shunt and current sensor resistors have little or no response to "X_T" until well up into the megahertz range. Therefore, the effect of reactance can largely be ignored.

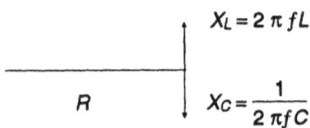

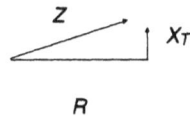

Figure 1.2   Figure 1.3

Thermal EMF

Dissimilar metals, in contact with each other, produce a small voltage. This voltage is variable with temperature and is therefore called a "Thermal EMF" or thermocouple effect. The rate of change of voltage with temperature from an intermetallic junction is a function of the metallic combination and the sense of the voltage produced is either positive or negative depending on which side of the combination is being considered the input. Virtually all resistors have intermetallic combinations and it is presumed will eventually be connected to copper as a final intermetallic junction. Hence copper is the typical reference metal. Figure 1.4 is a brief synopsis of the thermal EMFs for various metals and alloys used in resistor construction vs. copper as the reference.

Metal/Alloy	Thermal EMF vs. Copper $\mu V/^{\circ}C$
Evanohm	+2
Cupron	-45
Manganin	-3
Zeranin	-1.3
Nickel	-22
Gold	+0.2
Silver	-0.2
Aluminum	-4

Figure 1.4   Thermal EMF of Selected Metals and Alloys

Thermal EMF is an important consideration in low value resistors used in DC circuits. Thermal EMF usually has no importance in AC circuitry. In particular, the importance in current sensors is obvious since the thermal EMF could be larger than the signal being discriminated.

As noted earlier, thermal EMFs have polarity and so, for example, from copper to manganin and back to copper through a particular resistor, one end is a $+3\mu V/^{\circ}C$ generator and the other end is a $-3\mu V/^{\circ}C$ generator. In the ideal situation of both ends of the resistor being at the same temperature, the thermal EMFs are self-cancelling. The manufacturer minimizes thermal EMF effects through the use of appropriate metals but it is up to the user to see that the product is installed in such a way that the resistor is not being heated nonuniformly by other components. Figure 1.5 is the equivalent circuit when both ends of the resistor are at the same temperature and Figure 1.6 is the equivalent circuit when one end is heated. Note that the 100μV measured IR drop of Figure 1.5 became 130μV apparent drop in Figure 1.6 when a temperature difference of 10˚C is allowed to exist across the resistor.

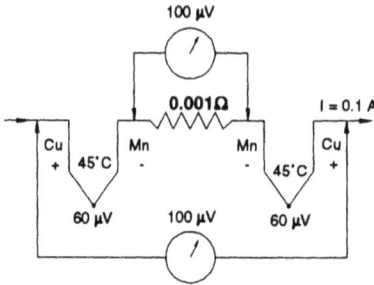 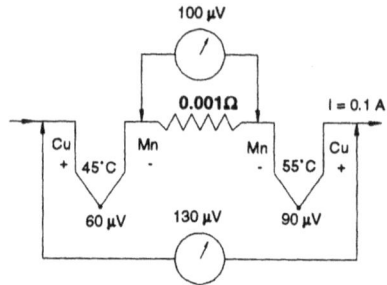

Figure 1.5   Figure 1.6

Noise

Noise is an unwanted AC signal from within the resistor. Two types of noise exist and are cause for limiting product usefulness under certain conditions.

THERMAL NOISE - Often called "Johnson Noise," thermal noise is due to the random motion of electrons in the resistive material which create small fluctuating potential differences across the terminations.

Thermal noise is characterized by a continuous frequency spectrum and its magnitude is independent of the material of the conductive element or its shape and varies only with temperature and resistance.

8

Thermal noise can be obtained from the formula:

$$E^2 = 4KTR \, (f2 - f1)$$

where:

K = BOLTZMANN'S constant (1.38 x 10-23 Joules/Degree Kelvin)
T = Absolute temperature (°C +273)
R = Resistance of the conductor
E = Root mean square value of the thermal noise voltage
(f2 - f1) = Bandwidth in Hertz.

The voltage developed by thermal agitation sets a limit on the smallest voltage that can be amplified without being lost in a background of noise.

CURRENT NOISE - Resistances composed of metal or metal alloys (wirewound, etc.) display the lowest combined noise level and can largely be ignored. However, resistances composed of conductive particles dispersed in an insulating matrix or films with imperfect lattice structure and nonconducting occlusions generate noise far in excess of the thermal agitation noise when a direct current is passed through the resistance. This type of noise arises from fluctuations in contact resistance between conducting sites within the matrix and is greater in higher values where the sites are fewer. It can also occur at poorly joined metal connections such as cold solder joints. The frequency spectrum of current noise is not continuous and appears in the lower audio frequency range. This may place a limitation on the use of thick film resistors under the circumstances of very small signal discrimination at mid-frequencies.

NOISE INDEX - The noise index is calculated from the formula:

$$db = 20 \, x \, \log_{10} \frac{Noise \ Voltage \ (over \ a \ 1 \ decade \ bandwidth)}{DC \ Voltage}$$

where noise voltage is the RMS current noise generated by the resistor and the DC voltage is the voltage arising out of the specified current flowing through the resistor at a specified temperature.

The index is usually expressed either as μV/V or in decibels of the ratio of voltages and these are interchangeable as shown in Figure 1.7.

db	μV/V
15	5.60
10	3.20
5	1.80
0	1.00
-5	0.56
-10	0.32
-15	0.18
-20	0.10
-25	0.056
-30	0.032
-35	0.018

Figure 1.7   Noise Conversion

Application Information

As with other types of components, the most important thing a designer must decide is which of the numerous types of resistors will be best for use in a particular application. Proper selection in its broadest sense is the first step in a reliable design.

When selecting a resistor, consider the following:
- AC requirements
- Stability
- Noise
- Power dissipation
- Environment
- Resistance value.

Actual resistance value is a function of tolerance, voltage coefficient, temperature coefficient, and drift with time.

The power rating is based upon ambient temperature and derating. Derating may be necessary because of environmental conditions.

Resistor compositions include carbon, film, and wirewound for fixed resistance units and cermet and conductive plastic for variable units. Metal foil can be used for both fixed and variable units.

All variable and fixed resistors, of the types widely used in electronic equipment, can be grouped into one of five general types according to construction type.

- **Composition type** - Made of a mixture of resistive material and a binder which are molded into the proper shape and value.
- **Film type** - Composed of a resistive film deposited on, or inside of, an insulating cylinder or filament.
- **Foil type** - Element of metal foil photo-etched to a specific pattern.
- **Wirewound type** - Made up of resistance wire, wound on an insulated form.
- **Nonwirewound type** - Constructed with a resistance element usually consisting of carbon, cermet, or conductive plastic deposited on a plastic insulating base.

These basic types differ from each other in size, cost, power rating, resistance range, and general characteristics. Some are better than others for particular purposes; no one type has all of the best characteristics. The choice among them, therefore, depends on the requirements, both initial and long-term; the environment in which they must exist; and numerous other factors which the designer must understand. The designer must realize that the summaries of the general characteristics are relative, not absolute, and that all the requirements of a particular application must be taken into consideration and compared with the advantages and drawbacks of each of the several types before a final choice is made.

Resistors should be mounted so resonance does not occur within the frequency spectrum of the vibration environment to which the resistors may be subjected. Resistor mounting for vibration should provide:

- The least tension or compression between the lead and body.
- The least excitation of the resistor in relation with any other surface.
- No bending or distortion of the resistor body.

Improper heat dissipation is the predominant contributing cause of failure for any resistor type; consequently, the lowest possible resistor surface temperature should be maintained. Figure 1.8 illustrates the manner in which heat is dissipated from fixed resistors in free air.

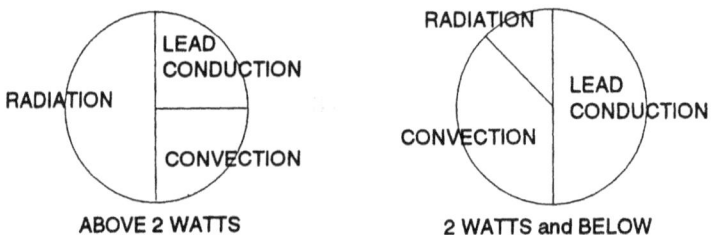

Figure 1.8 Heat Dissipation Under Room Conditions

The intensity of radiated heat varies inversely with the square of the distance from the resistor. Maintaining maximum distance between heat-generating components serves to reduce cross-radiation heating effects and promotes better convection by increasing air flow. For optimum cooling without a heat sink, small resistors should have large diameter leads of minimum length terminating in tiepoints of sufficient mass to act as heat sinks. All resistors have a maximum surface temperature which should never be exceeded. Any temperature beyond maximum can cause the resistor to malfunction. Resistors should be mounted so that there are no abnormal hot-spots on the resistor surface. When mounted, the resistors should not come in contact with heat-insulating surfaces.

Resistors that are crowded together and come into contact with each other can provide leakage paths (even well insulated parts) for external current passage. This can change the resultant resistance in the circuit. Moisture traps and dirt traps are easily formed by crowding. Moisture and dirt eventually form corrosive materials which can deteriorate the resistors and other electronic parts. Moisture can accumulate around dirt even in an atmosphere of normal humidity. Planning should be done to eliminate crowding of parts. Proper space utilization of electronic parts can reduce the package size and still provide adequate spacing of parts.

The following is a guide for resistor mounting:

- Maintain lead length to a minimum. The mass of the connection point acts as a heat sink. (NOTE: Where low temperatures are present, leads should be offset (bent slightly) to allow for thermal contraction.
- Close tolerance and low-value resistors require special precautions (i.e., short leads and good soldering techniques) since the resistance of the leads and the wiring may be as much as several percent of the resistance of the resistor.
- Maintain maximum spacing between power resistors.
- For resistors mounted in series, consider the heat being conducted through the leads to the next resistor.
- Large power units should be mounted to the chassis.
- Do not mount high-power units directly on terminal boards or printed circuits.

- To provide for the most efficient operation and even heat distribution, power resistors should be mounted in a horizontal position.
- Select mounting materials that will not char and can withstand strain due to expansion.
- Consider proximity to other heat sources as well as self-heat.
- Consider levels of shock and vibration to be encountered. Where large body mass is present, the body should be restrained from movement.

Large resistors that are not provided with some adequate means of mounting should not be considered. Under conditions of vibration or shock, lead failure can occur, and the larger the mass supported by the leads the more probable a failure will occur. Even when vibration or shock will not be a serious problem, ease of assembly and replaceability suggest that large components be mounted individually.

The body of the resistor must be sufficiently strong to withstand any handling it is likely to get. Through workmanship and packaging requirements, it should be shown by the manufacturer that their components will not crack, chip, or break in transit, on the shelf, or in the normal assembly process.

All resistors intended for use in reliable electronic equipment must be protected by an insulating coating. Sometimes this is a molded phenolic case, epoxy coating, or a ceramic glass sleeve. Wirewound power resistors use various cement and vitreous enamel coatings to protect the windings and to insulate and provide moisture barriers. Not all of the coatings and insulators applied to commercial resistors are satisfactory for extreme variations in ambient conditions. This information can be found in the construction materials list from the manufacturers published data.

In the establishment of ratings for resistors, the design engineer has considered the mechanical design of the product. The ambient conditions in which the resistor must operate determines to a large degree the power rating and mechanical construction of the resistor if long life under extreme conditions is desired.

A very important question in the application of resistors is how hot will they get in service. In a product the heat in a resistor comes from several sources:

- Self-generated heat, that can be easily calculated.
- The heat that the resistor receives from other resistors or other heat-producing components in the same immediate vicinity by radiation, and is not so easily calculated.
- The ambient temperature of the surrounding medium.

The important thing to remember is that under these conditions each resistor will be heated more than I^2R would suggest; when much heat is produced, as in stacked wirewound resistors, the design engineer would do well not to freeze the design until measurement of a typical assembly with power on to see just how hot the resistors get. The same is true of the extra heating given the resistors by convection. This is another way of saying that high-ambient temperature will reduce the actual power rating of the resistor by reducing permissible temperature rise. The designer must realize also that the heat being produced by "hot" resistors can injure other components. This is a very important point to remember; capacitors, diodes, and other resistors usually do not fail immediately when overheated. The effect of too much heat is a deteriorating one, weakening the component until at a later date it will unexpectedly fail. It is very easy to put a "heat bomb" in a product that will not go off in normal production testing but will do so when the product gets into service and is being relied on to do its operation. It is also very easy to eliminate such troubles by strict and thoughtful attention to the problem of heating.

Most of the manufacturer's tests are performed at ambient atmospheric pressure. This fact should be considered when resistors are used for high-altitude conditions or other pressurized environments.

Flammability should be taken into consideration in all applications.

The manufacturer's failure rate figure (in percent per 1,000 hours) is usually based on the single parameter of load life test results only.

The resistance value is initially determined by the circuit requirements, and may seem a trivial thing to mention. However, most resistor calculations that are made without reference to available resistors come out to a resistance value that is not standard. The design engineer should be aware of the standard resistance values that are available from manufacturers. These differ somewhat with the various types of resistors. It is usually a fairly simple thing to bring the exact calculated value in line with a standard value. In the case where this cannot be done, a parallel or series combination of resistors can usually be used.

The design engineer should also remember that the resistance value of the resistor that gets into the physical circuit will differ from the value he has stated on his circuit schematic, and that this difference will change as time goes by. The purchased tolerance of the resister to be used will allow it to differ from the nominal stated value, depending of the type of resistor specified. Furthermore, the temperature at which the resistor works, the voltage across it, and the environment in which it lives will affect the actual value at particular times.

For example, the designer should allow for a possible variation of ±15 percent from the nominal value of a purchased ±5 percent composition resistor, if one expects the circuit to continue to operate satisfactorily over a very long time under moderate ambient conditions. Such a figure is a rule of thumb, based on many tests, and many resistors will remain much nearer their starting value; but if many are used, chance will ensure that some will go near this limit. A similar figure can be deduced from each variety of resistor used.

The minimum required power rating of a resistor is another factor that is initially set by the circuit usage, but is affected by other conditions of use. The power rating is based on the "hot-spot" temperature the resistor will withstand, while still meeting its other requirements of resistance variation, accuracy, and life.

Self-generated heat in a resistor is calculated as $P=I^2R$. This figure, in any circuit, must be less than the actual power rating of the resistor used. It is the usual practice to calculate this value and to use the next larger power rating available. This calculation should, however, be considered only as a first approximation of the actual rating to be used.

The power rating of a resistor is based on a certain temperature rise from an ambient temperature of a certain value. If the ambient temperature is greater than this value, the amount of heat that the resistor can dissipate is correspondingly reduced, and therefore it must be derated because of temperature. Most manufacturer's data books contain derating curves to be applied to the resistors.

Because of the temperature coefficient of resistance that all resistors possess, a resistor which is expected to remain near its specified value under conditions of operation must remain cool. For this reason, all resistors designated as "accurate" are very much larger physically for a certain power rating than are ordinary "nonaccurate" resistors. In general, any resistor, "accurate" or not, must be derated to remain very near its original specified valued when it is being operated.

If especially long life is required of a resistor, particularly when "life" means remaining within a certain limit of resistance drift, it is usually necessary to derate the resistor, even if ambient conditions are moderate and accuracy by itself is not important. A good rule to follow when choosing a resistor size for equipment that must operate for many thousands of hours is to derate it to one half of its nominal power rating. Thus, if the self-generated heat in the resistor is 1/3 watt, do not use a 1/2 watt resistor, but rather a 1 watt size. This will automatically keep the resistor cooler, will reduce the long-term drift, and will reduce the effect of the temperature coefficient. In products that need not live so long and must be small in size, this rule may be impractical, and the designer should adjust the dependence on rules to the circumstances at hand. A "cool" resistor will generally last longer than a "hot" one, and can absorb transient overloads that might permanently damage a "hot" resistor.

When a resistor is used in circuits where power is drawn intermittently or in pulses, the actual power dissipated with safety during the pulses can sometimes be much more than the maximum rating of the resistor. For short pulses, the actual heating is determined by the duty factor and the peak power dissipated. Before approving such a resistor application, however, the designer should be sure:

- That the maximum voltage applied to the resistor during the pulses is never greater than the permissible maximum voltage for the resistor being used.
- That the circuit cannot fail in such a way that continuous excessive power can be drawn through the resistor and cause it to fail also.

- That the average power being drawn is well within the agreed upon rating of the resistor.
- That continuous steep wave fronts applied to the resistor do not cause any unexpected troubles.

For most resistors the lower the resistance value, the less total impedance it exhibits at high frequency. Resistors are not generally tested for total impedance at frequencies above 120 Hertz. Therefore, this characteristic is not controlled. The dominating conditions for good high-frequency resistor performance are geometric considerations and minimum dielectric losses. For the best high-frequency performance, the ratio of resistor length to the cross sectional area should be a maximum. Dielectric losses are kept low by proper choice of the resistor base material, and when dielectric binders are used, their total mass is kept to a minimum. The following is a discussion of the high-frequency merits of the major resistor types:

- **Carbon-composition:** This type exhibits little change in effective DC resistance up to frequencies of about 100 kHz. Resistance values above 0.3 megohm start to decrease in resistance at approximately 100 kHz. Above frequencies of 1 MHz, all resistance values exhibit some decreased resistance.
- **Metal:** Metal-type resistors have the best high-frequency performance. The effective DC resistance for most resistance values remains fairly constant up to 100 MHz and decreases at higher frequencies. In general, the higher the resistance value the greater the effect of frequency.
- **Wirewound:** Wirewound resistors have inductive and capacitive effects and are unsuited for use above 50 kHz, even when specially wound to reduce the inductance and capacitance. Wirewound resistors usually exhibit an increase in resistance with high frequencies because of "skin" effect.

Much trouble during the life of the product can be eliminated if the designer can be sure that the resistors specified for the circuit are soundly constructed and proper equipment assembly techniques are utilized. The resistor types used should provide a great measure of this assurance and, in general, assure a uniform quality of workmanship.

The connection between the resistor element itself and the pigtails of leads that connect it into the circuit must be so good that no possible combination of conditions met in the proposed operation can cause an intermittent connection. When resistors are handled in automatic assembly machines, this precaution is particularly important.

There are assembly techniques that affect resistor reliability. Resistors should never be overheated by excessive soldering-iron applications, and the resistor leads should not be abraded by assembly tools. No normal soldering practice, either manual or dip soldering, should damage the resistor physically or change its resistance value appreciably.

Moisture is the greatest enemy of components and electronic equipment. Usually a resistor will keep itself dry because of its own self-generated heat; this is, of course, only true when the product is turned on. If the product must stand for long periods under humid conditions without power applied, the designer should determine whether the circuits will operate with resistance values which have changed from the "hot" condition, and whether the retrace of the resistance value during the warm-up period will allow the product to work satisfactorily during this period. If it will not, the designer must see that a resistor adequately protected against moisture absorption is used. The resistor cannot be blamed for performing improperly if it is not designed for the conditions of use. It is therefore up to the designer to analyze what is needed and to provide a resistor to meet these conditions.

To summarize:

- Select a resistor for each circuit application from the lists of standard types and values. (See Appendix B)
- Be sure that the circuit being designed will work with any resistor whose resistance value is within the limits set by tolerance plus voltage coefficient plus temperature coefficient plus drift with time. Failure to take these precautions can possibly mean that when the product is produced in quantity, there may be some circuits that will not work under extreme conditions.
- Various initial tolerances are available depending on the type of resistor. It should be remembered that initial accuracies become meaningless if the inherent stability of the resistor does not support the initial accuracy.

- During shelf life, as well as during operational life, any characteristic (i.e., resistance, inductance, power rating, dielectric strength, size, etc.,) of any part may change value due to stresses caused by environmental changes of temperature, humidity, pressure, vibration, etc. Changes of characteristics caused by environmental stresses may be linear or nonlinear, reversible or nonreversible (permanent), or combinations thereof. Where a characteristic of the part undergoes a linear change during environmental stress, and the change reverses itself linearly when the environmental stress is removed so that the characteristic returns to its normal value, this rate of change in characteristic value (per unit change in stress value) is designated (x) coefficient, and is usually expressed in percent or ppm/°C.

Chapter 2

Composition Resistors

Carbon-composition units have a resistive element that is molded from carbon powder that has been mixed with a phenolic binder to form a uniform resistive body. That body, molded with end leads, is a general purpose resistor capable of withstanding temperature and electrical transient shocks. The resistors are covered by a molded or conformal coating jacket which is primarily intended to provide an adequate moisture barrier for the resistance element, as well as mechanical protection and strength. The carbon-composition resistor is used in applications where initial tolerance need not be closer than ±5 percent with long term stability no better than ±20 percent.

Carbon elements are susceptible to moisture absorption and such moisture absorption can cause the resistance to change as much as 20 percent. That resistance shift can be reversed if the device is baked at high temperatures (100°C) for 92 to 100 hours.

General characteristics of fixed, composition resistors:

- Nominal minimum resistance tolerance available for fixed, composition resistors is ±5 percent. Combined effects of climate and operation on unsealed types may raise this tolerance to ±15 percent from the low value (i.e., aging, pressure, temperature, humidity, voltage gradient, etc.).

- High-voltage gradients will produce resistance changes during operation.
- High "Johnson" noise levels at resistances above 1 megohm preclude use in critical circuits of higher sensitivity.
- RF will produce end-to-end shunted capacitive effects because of short resistor bodies and small internal distances between both ends.
- Operation at VHF or higher frequency reduces effective resistance due to losses in the dielectric (the so-called "Boella" effect).
- Exposure to humidity may have two effects on the resistance value: 1) Surface moisture may result in leakage paths which will lower the resistance values or 2) absorption of moisture into the element may increase the resistance. This phenomenon is more noticeable in higher ranges since it depends upon the resistance value.
- The resistance temperature characteristic is the highest for general purpose resistor styles.

Variable composition resistors have a composition resistance element shaped in an arc, and a contact bearing uniformly thereon, so that a change of resistance is produced between the terminal of the contact and the terminal at either end of the resistance element when the operating shaft is turned. The construction of the element is usually a molded type which is a one-piece unit containing the resistance material, terminals, face plate, and the bushings. A heat bonding of the element and form is then performed. The element is then contained in an enclosure which provides environmental and mechanical protection. For variable resistors, one problem is that the carbon element requires a high contact force to ensure that any variation in the contact resistance remains within acceptable limits. That results in high shaft-torque and poor adjustability.

A linear resistance taper is one having a constant change of resistance with angular rotation.

A nonlinear resistance taper has a variation or lack of constancy in the change of resistance with angular rotation.

Application Information

Use fixed insulated resistors for general purpose resistor applications where the initial tolerance needs to be no closer than ±5 percent and long term stability needs to be no better than ±15 percent under fully rated operating conditions.

Use variable resistors where initial setting stability is not critical and long-term stability needs to be no better than ±20 percent.

Variable composition resistors are suitable for rheostat or potentiometer applications where stability and high precision are not required, and are capable of withstanding acceleration, shock, and high-frequency vibration. They are most useful in circuitry where high resistance values and low power dissipation are encountered in controlling volume, bias, tone, voltage output, and pulse width. Composition, variable resistors are useful only up to the low radio frequency ranges.

When considering variable composition resistors for potentiometer applications, it is necessary to bear in mind the fact that the load current as well as the "bleeder" current will be flowing through a part of the resistor and will contribute to the heating effect.

Variable carbon-composition resistors should not be used at high potentials to ground (greater than 200 volts peak), unless supplementary insulation is provided.

The noise level of variable composition resistors is quite high compared to other types of resistors. Thermal and mechanical noise level will normally decrease over the life of the resistor.

Consideration must be given to the resistor's wattage rating. This is based on the materials used and is controlled by specifying a maximum "hot-spot" temperature. The amount of dissipation that can be developed in a resistor body at the maximum "hot-spot" temperature depends upon how effectively the dissipated energy is carried away and therefore it is also a direct function of the ambient temperature. To be operated continuously at full rating, the resistor must be connected to an adequate heat sink, which means a sufficient amount of length of lead connected to average size solder terminals with no other dissipative parts connected to the same terminals or mounted closer than one diameter. Appropriate derating must be imposed at elevated temperatures. Power dissipation capabilities of a resistor are usually lower when mounted in a product than under test conditions. Most of the generated heat is carried away by the resistor leads; therefore, when two resistors are connected to the same terminal, wattage ratings would be decreased approximately 25 percent. Close proximity of one resistor to another, or to any other heat generating part, further reduces the wattage rating. Conformal coatings and encapsulating materials are poor heat conductors. When resistors are packaged in this manner, exercise caution in selection of the power rating.

When a resistor is to be used in a circuit where the surrounding temperature is higher than the base temperature used for the power rating, a correction factor must be applied to the wattage rating so as not to overload the resistor. This correction factor may be taken from the curve known as the derating curve for high ambient temperature.

For optimum performance, two "rules of thumb" have been practice in the industry for composition resistors:

- After the anticipated maximum ambient temperature has been determined, a safety factor of 2 is applied to the wattage.
- Wattage is adjusted so that the "hot-spot" temperature does not exceed the following for the particular style:
  120°C - 1/8 watt and 1/4 watt.
  100°C - 1/2 watt, 1 watt, and 2 watt.

NOTE: It is recommended that either of the above techniques be considered in the application of composition resistors.

When composition resistors are used under low-duty-cycle pulse conditions, the maximum permissible operating voltage is limited by breakdown rather than by heating. In such applications the peak value of the pulse should not exceed $\sqrt{2}$ times the rated rms continuous working voltage for the type used. If the pulses are of sufficient duration to raise the resistor's temperature excessively, the resistor must be derated even though the interval between pulses may be long enough to make the average heating small. In general, the above procedure must be used with caution if it permits the peak power to be more than approximately 30 to 40 times the normal power rating.

Thermal agitation or "Johnson" noise and resistance fluctuation or carbon noise, present only when current is flowing, are characteristic of carbon composition resistors. Use of these resistors in low level high-resistance (one megohm or more) circuits should be avoided. Noise which can be expected is approximately 3 to 10 microvolts per volt. A film or wirewound resistor will usually yield more satisfactory results.

When exposed to humid atmosphere while dissipating less than 10 percent of rated power (including shelf storage, equipment nonoperating, and shipping conditions), resistance values may change 15 percent.

The fact that there are voltage limits in the applications of fixed composition resistors is often overlooked. These maximum permissible voltages, which are imposed because of insulation breakdown problems, must be taken into consideration in addition to the limitations of power dissipation.

When used in high frequency circuits (1 MHz and above), the effective resistance will decrease as a result of dielectric losses and shunt capacity (both end-to-end and distributed capacity to mounting surface). High frequency characteristics of carbon composition resistors are not controlled and hence are subject to change without notice. Typical examples of changes in effective resistance are:

- At 1 MHz, a 1/2W, 100k ohm resistor measures 90% of DC value.
- At 10 MHz, a 1/2W, 100k ohm resistor measures 55% of DC value.
- At 10 MHz, a 2W, 1M ohm resistor measures 15% of DC value.

Voltage coefficient should be considered for use in circuits which are sensitive to this parameter. When voltage is applied to carbon composition resistors, resistance values may change by up to 2 percent.

The resistance-temperature variations of carbon composition resistors cannot be defined by a temperature coefficient since the variation is not only nonlinear but is a different shape for different resistance values.

Shelf life, in general, of fixed resistors can exhibit resistance variations as high as ±15 percent due to moisture and temperature effects. When a closer life tolerance or higher accuracy is needed, fixed film resistors should be used.

Care should be taken in soldering resistors, since all properties of a composition resistor may be seriously affected when soldering irons are applied too closely to the resistor's body or terminals for too long a period of time. The length of lead left between the resistor body and the soldering point should not be less than 1/4 inch. Heat-dissipating clamps should be used, if necessary, when soldering resistors in close quarters. In general, if it is necessary to unsolder a resistor to make a circuit change or in maintenance, the resistor should be discarded and a new one used.

For conditioning purposes and to remove moisture, composition resistors are baked at 100°C for 92 to 100 hours.

Failures are considered to be opens, shorts, or radical departures from initial characteristics occurring in an unpredictable manner and in too short a period of time to permit detection through normal preventive maintenance. The failure rate factors based on "catastrophic failures" will differ from failure rates established on "parametric failures" of ±15 percent change in resistance to be expected at 0 to 10,000 hours of life at rated conditions.

Chapter 3

Film Resistors

Metal film devices are used in applications requiring higher stability and precision than available from carbon devices. In addition, metal film resistors should be used in applications where AC is present. Operation is satisfactory from DC to the megahertz range. Metal film units have low temperature coefficients and suffer little degradation to ambient temperatures of 125°C and higher. Film resistors can be classified according to the techniques used in their manufacture.

One technique is vacuum deposition, which is also known as evaporated metal film. In it, a nickel-chromium alloy is superheated in a vacuum. The alloy vaporizes and is deposited on a ceramic substrate. Small quantities of contaminants, called dopants, are used to control resistor characteristics such as resistance range. Those resistors are used in applications that require an extreme degree of precision.

In sputtering, a nichrome target is heated and bombarded by agron atoms. That results in metal atoms being knocked off and deposited on a substrate. Resistors manufactured using this sputtering technique are also suitable for applications that require a high degree of precision.

In metal-oxide deposition, a chemical vapor is used to deposit a tin-oxide film onto a glass substrate. That technique is used to produce resistors for general-purpose, semi-precision, and precision applications.

Typical thin-film resistors are sputtered tantalum nitride, deposited chromium cobalt, or nichrome, on a substrate. Substrates of alumina, sapphire, glass, quartz, beryllia, or silicon are used.

Thin-film resistor networks are also available; these are housed in Dual Inline Packages (DIPs), Single Inline Packages (SIPs), or flat pack configurations. In these resistors, the element consists of a film element on a ceramic substrate. The element is formed either by deposition of a vaporized metal or the printing of a metal and glass combination paste which has then been fired at a high temperature. Resistance elements are generally rectangular in shape and calibrated to the proper resistance value by trimming the element by abrasion or a laser beam. After calibration, the resistance element is protected by an enclosure or a coating of insulating, moisture-resistance material (usually epoxy or a silicone).

Chip resistors are uncased, leadless resistance element chip devices that possess a high degree of stability with respect to time, under severe environmental conditions.

In individual chip resistors, the terminals used may be either surface or wrap-around types. Wrap-around terminals wrap around the side of the substrate allowing connections to the underside. Terminals of solder, silver over nickel, paladium-silver, platinum, or platinum gold plating are available. Trimming of the resistor is done either mechanically or by using a laser.

In thick-film resistors, a ceramic substrate is coated (silk screened - a mechanized stenciling process) with a glass-metal material and then fired (to cure it) at a high temperature. The glass-metal materials include nichrome, silver palladium, platium, ruthenium, rhodium, gold, and a tantalum-modified tin oxide. That film is up to 100 times thicker than evaporated or sputtered metal film (greater than 0.0001 inches thick) and is used in applications requiring high power density or the capability of surviving power spikes or overloads. Those units are suitable for some precision applications, but not those requiring an extremely high degree of precision.

Bulk metal resistors are made when metal foil is laminated to a substrate and chemically etched to produce a conductive path. The flat element is used exclusively for high-precision applications and has tight tolerances and an excellent temperature coefficient.

A fixed film resistor element consists of a film-type resistance element which has been formed on a substrate. The element is spiraled to achieve ranges in resistance value, and after lead attachment, the element is coated to protect it from moisture or other detrimental environmental conditions.

Carbon-film resistors were introduced to perform the same basic functions as carbon-composition resistors, but at a lower price. Just like composition types, they lack the ability to withstand transient voltage spikes and have a poor temperature coefficient.

An axial-lead, carbon-film resistor is made by screening carbon based resistive inks on a ceramic rod and then firing the assembly. Alternate techniques include depositing pure carbon by cracking a hydrocarbon gas or by depositing a nickel film for resistor values of less than 10 ohms. The resistive element may also be sprayed on, applied with a transfer wheel, or dipped on. The rod is then cut to size, leaded end caps are attached, and the unit is trimmed to a precise value. The resistor is then coated with an insulating material. Carbon-film resistors are available in the same resis- tance values as carbon-composition units and have a typical tolerance of ±5 percent.

General characteristics of fixed film resistors:

- Low tolerance; high stability; low environmental changes; low temperature coefficient; spacing and weight saving; low noise.
- Nominal minimum resistance tolerance available is ±0.1 percent for fixed film resistors; and for the resistor networks, the nominal minimum resistance tolerance available is ±1.0 percent.
- Maximum practical full-power operating temperature should not be exceeded. Resistor networks and resistor chips are continuously derated.

- Operation at RF (above 100 MHz) may produce inductive effects on spiral-cut type fixed film resistors and capacitive effects on the resistor networks.
- The resistance temperature characteristic is fairly low for thick film types and very low for metal film type. The resistance temperature characteristic is fairly low for resistor networks and resistor chips.

Application Information

Fixed film-type resistors are used in applications demanding better stability, tolerance, and temperature coefficient requirements than carbon composition types. For applications requiring greater precision and tighter tolerances, the use of metal film or wirewound resistors is recommended.

Fixed film (high stability) resistors are used in circuits requiring higher stability than provided by composition resistors and where AC frequency requirements are critical. Operation is satisfactory from DC to 100 megahertz. Metal films are characterized by low temperature coefficients and are usable for ambient temperatures of 125°C or higher with little degradation.

Fixed metal film resistors are designed for use in critical circuitry where high stability, long life, reliable operation, and accuracy are of prime importance. They are particularly desirable for use in circuits where high frequencies preclude the use of other types of resistors. Some of the applications for which these metal film type resistors are especially suited are as follows:

- High-frequency, tuned circuit loaders
- Television side-band filters
- Rhombic antenna terminators
- Radar pulse equipment
- Impedance metering bridges
- Standing wave-ratio meters.

Thin-film resistors are highly stable, have low-noise characteristics, and have a very low temperature coefficient. They are used in:

- Digital multimeters
- Precision voltage-dividers
- Attenuators
- A/D and D/A circuits
- Current-summing applications.

Chip film resistors are primarily intended for incorporation into hybrid microelectronic circuits. They are designed for use in critical circuitry where stability, long life, reliable operation, and accuracy are of prime importance.

Operation of resistor chips under extreme high or low ambient conditions may cause permanent or temporary changes in resistance sufficient to exceed their initial tolerances.

Because of the very small size of the resistance elements and connecting circuits, there are maximum permissible voltages which are imposed.

Film networks are designed for use in critical circuitry where stability, long life, reliable operation, and accuracy are of prime importance. They are particularly desirable for use where miniaturization is important and ease of assembly is desired. They are useful where a number of resistors of the same resistance value are required in the circuit.

Resistors within a network have a power rating based on continuous, full-load operation at an ambient temperature. A power rating is given for each resistor within the network and a power rating is given for the total network package. The package power is usually not equal to the individual resistor power rating times the number of resistors within the network.

Because all the electrical energy dissipated by a resistor is converted into heat energy, temperature of the surrounding area is an influencing factor when selecting a particular resistor for a specific application. The power rating of these resistors is based on operation at specific temperatures; however, in actual use, the resistor may not be operating at these temperatures. When the desired characteristic and the anticipated maximum ambient temperatures have been determined, a safety factor of 2, applied to the wattage, is recommended in order to insure the selection of a resistor having an adequate wattage-dissipation potential for optimum performance.

Film resistors have semi-precision characteristics and small sizes. Good stability makes them desirable in most electronic circuits.

Metal film resistors are essentially unaffected by moisture.

When used in high frequency circuits (200 MHz and above), the effective resistance will decrease as a result of shunt capacity (both end-to-end and distributed capacity to mounting surface).

When metal film resistors are used in low duty-cycle pulse circuits, peak voltage should not exceed 1.4 times the Rated Continuous Working Voltage (RCWV). However, if the duty-cycle is high or the pulse width is appreciable, even though average power is within ratings, the instantaneous temperature rise may be excessive, requiring a resistor of higher wattage rating. Peak power dissipation should not exceed four times the maximum rating of the resistor under any condition.

Noise output for metal film resistors is a negligible quantity. In applications where noise is an important factor, fixed film resistors are superior to composition types.

Under conditions of severe shock or vibration (or a combination of both), resistors should be mounted in such a fashion that the body of the resistor is restrained from movement with respect to the mounting base. It should be noted that if clamps are used, certain electrical characteristics of the resistor will be altered. The heat-dissipating qualities of the resistor will be enhanced or retarded depending on whether the clamping material is a good or poor heat conductor.

Under relatively low humidity conditions, some types of chip film resistors, particularly those with small dimensions and high sheet resistivity materials, are prone to sudden significant changes in resistance (usually reductions in value) and to changes in temperature coefficient of resistance as a result of discharge of static charges built up on associated objects during handling, packaging, or shipment. Substitution of more suitable implements and materials can help minimize this problem. For example, use of cotton gloves, static eliminator devices, air humidifiers, and operator and work bench grounding systems can reduce static build-up during handling. Means of alleviating static problems during shipment include elimination of loose packaging of resistors and use of metal foil and anti-static (partly conducting) plastic packaging materials.

Chapter 4

Foil Resistors

Metal foil resistor technology, invented by physicist Dr. Felix Zandman in 1962, out-performs all other resistor technologies available today for applications that require ultra precision and stability. Metal foil resistors could be called the "perfect" resistor. A typical metal foil resistor application is in a satellite which requires stable positioning circuitry to function through temperature extremes. To be able to achieve this, the only solution is the use of extremely low temperature coefficient of resistance (TCR) resistors.

Metal foil resistor's superior performance is the result of the chip resistor element. The chip element consists of metal foil on a ceramic substrate. The metal foil has been photo-fabricated to a specific resistance pattern. The more homogeneous the resistive material, the better the retention of all of the original characteristics. Thus, metal foil that is formed mechanically from the original alloy exhibits significantly better electrical characteristics than evaporated and sputtered metal films (both which have been molecularly disassembled and reassembled). This combination of materials (metal foil and ceramic substrate) provides a resistor element with characteristics that are unavailable in other technologies. Various chip sizes and resistor configurations are used to provide the variations in power, size, and other operating specifications for various applications.

Leads make contact with the resistor element and then the unit is encased for protection and ease of handling. Conformally coated, molded case, surface mount, and hermetically sealed units are available.

Wire-bondable chips are available when the resistor chip element is used in a hybrid circuit.

The tolerance of metal foil resistors can be as precise as ±0.001% (10 ppm) by selectively trimming various adjusting points that have been designed into the photo-etched pattern of the resistive element. They provide predictable step increases in resistance to the desired tolerance level. Trimming the pattern at one of these adjusting points forces the current to seek another longer path, thus raising the resistance value of the element by a specific percentage. In the fine adjustment areas, trimming affects the final resistance value by smaller and smaller amounts down to 0.001% and finally 0.0005% (5 ppm). This is the trimming resolution of commercial metal foil resistors.

A system or a device or one particular circuit element must perform for a specified period of time and at the end of that service period it must still be performing within specifications. During its useful life it may have been subjected to some hostile service conditions and it is no longer within purchased tolerance. One reason for specifying a tighter purchased tolerance than the end-of-life error budget tolerance is to allow for service shifts. Savings can be achieved by using tighter tolerance resistors and using the error budget on more expensive components.

Foil TCR - Two predictable and opposing physical phenomena within the composite structure of the resistance alloy and its substrate are the key to the low TCR capability of metal foil resistors. They are:
- Resistivity of the alloy changes predictably with temperature. (The resistance of the foil, a nickel-chrome alloy, *increases* as temperature increases.)
- The coefficient of thermal expansion of the alloy (α_a) is larger than the ceramic (α_c), $\alpha_a > \alpha_c$, resulting in a compressive strain on the resistive alloy when temperature increases. As a consequence, the resistance of the foil *decreases* due to compressive strain caused by the temperature increase. (Kelvin's Law).

The resistive foil is chosen so that it's resistance predictably increases with a temperature increase. The two effects occur simultaneously with changes in temperature. When matching is good, the result is an unusually low and predictable TCR - essentially none or a very low change of resistance with temperature.

Nominal TCR is defined as the chord slopes of the relative change of resistance/temperature (RT) curve, expressed in ppm /°C. Slopes are usually defined from:

- 0° C to + 25° C and + 25° C to + 60° C ("Instrument" Range).
- – 55° C to + 25° C and + 25° C to + 125° C ("Military" Range).

These specified temperatures and the defined nominal TCR chord slopes apply to all resistance values including low value resistors down to 1 ohm. Note, however, that without using four terminals and Kelvin connection in low values, allowance for lead resistance and associated TCR may have to be made.

Metal foil provides consistent chord slopes for all values and lots. This is a major improvement in TCR over other resistors. Metal foil resistors are used for applications requiring low TCR characteristics.

Foil technology has advanced through the years and TCR improvements have been achieved. Alloys have been developed that give positive or negative hyperbolic response to temperature. A method of achieving virtually zero TCR has been developed resulting in little or no response to temperature.

Tracking - When two or more resistors share a common substrate, improved TCR tracking performance is possible. Differential TCR tracking is the degree to which two resistor temperature coefficients of resistance remain within a specific variation track during temperature excursions. Depending on technology differences, some resistors may increase in value with an increase in temperature (+ TCR) while other resistors will decrease in value with an increase in temperature (- TCR), or they may not change in value at the same rate. Other temperature effects, such as self-heating due to the application of power can add to the ambient temperature effects. An example of these effects can be seen when two resistors with different TCR characteristics are used around an operational amplifier. The amplification ratio will be affected by the differential TCR of the resistors and will be compounded by the self-heating effects of the I^2R differences of the feedback vs. the sense resistor. Good design practice would be to use fundamentally low TCR networks in this application since this would minimize both varying temperature and self-heating effects. This could not be accomplished with high TCR resistors, even with good tracking.

Matching - This term describes the degree to which resistor characteristics remain within a specified variation at a specified temperature. The "match" or resistance ratio of a network of different values (not 1:1), can also be affected by TCR and self-heating effects. The self-heating effect on "matching" is often overlooked. Even though the initial match is tight and good TCR "tracking" is exhibited, the same current flow through the resistors will produce power dissipation differences (I^2R self-heating) and induce ratio changes proportional to the absolute TCR. Therefore, the lower the absolute TCR, the less the match will be affected. Good TCR tracking must be accompanied by very low absolute TCR to avoid temperature gradient effects due to ambient temperature differences or self-heating differences from one resistor to the other.

Load-Life Stability - This characteristic is most relied upon to demonstrate a resistor's long term stability. The military requires testing to 10,000 hours with limits on the amount of change in the resistor. Precision metal foil resistors have the tightest allowable limits. Whether military or not, the load-life stability of foil resistors is unparalleled and long term stability is assured. The reason foil resistors are so stable has to do with the materials of construction (metal foil and a ceramic high-purity alumina substrate).

The two parameters which must be mentioned together, power rating and ambient temperature, can be joined into one single parameter for a given style of resistor. If the steady-state temperature rise can be established, it can be added to the ambient temperature, and the sum will represent the combined (load induced plus ambient) temperature. The combined temperature comprises the effect of power induced temperature rise and the ambient temperature. Reducing the ambient temperature has a marked effect on load-life results.

Ratio - The ratio tolerances available in a network are variable and dependent on resistor technology. Ratios in a metal foil network formed in a common package have a better chance of holding tight tolerances than those formed in discrete matched sets by sorting. Hermetic foil networks offer the tightest ratio tolerances available in any technology. Theoretically, such low ratios as these can be made initially with other resistor technologies, however, less stable products will not hold these tight ratio tolerances over time with applied stresses of voltage and temperature.

Resistors in network form are expected to match at more than ambient temperature. For example, throughout the service life of the equipment, the resistors around the operational amplifier are required to match (to hold ratio) even though the dissipation in the feedback resistor is different than that in the sense resistor, causing one resistor to be at a higher temperature than the other. This is called matching under power. If environmental stresses cause one resistor to drift (permanent delta R's) more than its counterpart, the ratio changes over a period of time and this change can be significant. This is called matching with time. Foil resistors used in network form offer the best combination of temperature-load-time matching. The factors that contribute to this are:

- Fundamentally low TCR.
- Very small drift with load over time.
- Common behavior - all parts move in the same direction with temperature, load and time.

Speed - The equivalent circuit of a resistor combines a resistor in series with an inductance and in parallel with a capacitance. Resistors can perform like an R/C circuit or filter or inductor depending on the geometry. In spiraled and wirewound resistors these reactances are created by the loops and spaces formed by the spirals or turns of wire. Inter-loop capacitance increases with the number of loops or spirals. Current in adjacent loops travels in the same direction increasing inductance by mutual fields. The capacitance and inductance increase as the resistance value increases due to continually increasing the number of spirals or turns.

In planar resistors such as foil resistors, the geometry of the lines of the resistor patterns are intentionally designed to counteract these reactances. In the serpentine pattern of a planar resistor, the inter-loop capacitance reduces in series and the mutual inductance reduces due to current direction. The opposing directions of current prevents the build-up of mutual inductance and reduces the capacitive effects by placing the inter-conductor capacitances in series.

In pulse applications, these reactive distortions can result in a poor replication of the input. A pulse width of one nanosecond would be completely missed by a wirewound resistor while the foil resistor achieves full replication in the time allotted.

In frequency applications, these reactive distortions also cause changes in apparent resistance (impedance) with changes in frequency. Metal foil resistors exhibit very good response in the 100 ohm range out to 100 MHz and all values have good response out to 1 MHz. The performance curves for other types of resistors can be expected to demonstrate considerably more distortion (particularly wirewounds).

Noise - Resistors, depending on construction, can be a source of noise. This unintended signal addition is measurable and independent of the presence of a fundamental signal. As measurement instrumentation and circuitry become more demanding, noise (or unwanted signals) superimposed upon the fundamental signal become troublesome. Measurement instrumentation based on low-level signal inputs and high-gain amplification cannot tolerate microvolt level background noise when the signal being measured is itself in the microvolt range. Although audio circuitry, when signal purity is of utmost concern, is the most obvious use of noise-free components, other industries and technologies are equally concerned with this characteristic.

Resistors made of conductive particles in a nonconductive binder are the most likely to generate noise. In carbon composition and thick film resistors, conduction takes place at points of contact between the conductive particles within the binder matrix. Where these point-to-point contacts are made constitutes a high resistance conduction site which is the source for noise. These sites are sensitive to any distortion resulting from expansion mismatch, moisture swelling, mechanical strain, and voltage input levels. The response to these outside influences is an unwanted signal as the current finds its way through the matrix.

Resistors made of metal alloys, such as the metal foil resistor, are the least likely to be noise sources. Here the conduction is across the inter-granular boundaries of the alloy. The inter-granular current path from one or more metal crystals to another involves multiple and long current paths through the boundaries reducing the chance for noise generation.

In addition, the photo lithography and fabrication techniques employed in the manufacture of foil resistors results in more uniform current paths than found in some other resistor constructions. Spiraled resistors, for example, have more geometric variations that contribute to insertion of noise signals. Foil resistors have the lowest noise of any resistor technology, with the noise level being essentially immeasurable.

As mentioned earlier, resistors can change value due to applied voltage. The term used to describe the rate of change of resistance with changing voltage is known as "voltage coefficient." Resistors of different constructions have noticeably different voltage coefficients. In the extreme case, the effect in a carbon composition resistor is so noticeable that the resistance value varies greatly as a function of the applied voltage. Foil resistor elements are insensitive to voltage variation and the designer can count on foil resistors having the same resistance under varying circuit voltage level conditions. The inherent bulk property of the metal alloy provides a nonmeasurable voltage coefficient.

Thermal EMF - When a junction is formed by two dissimilar metals, and is heated, a voltage is generated due to the different levels of molecular activity within these metals. This electromotive force, induced by temperature, is called "Thermal EMF" and is usually measured in microvolts/°C.

In resistors, this thermal EMF is considered a parasitic effect interfering with pure resistance. It is often caused by the dissimilarity of the materials used in the resistor construction especially at the junction of the resistor element and the resistor lead materials. The thermal EMF performance of a resistor can be degraded by external temperature variations, dissymmetry of power distribution within the resistor element, and the dissimilarity of the molecular activity of the metals involved.

The level of voltage output is a function of the metals involved and the differential temperature. The voltage produced is either positive or negative depending on which side of the junction is considered as the reference point. Virtually all resistive products have these metallic junctions. It is presumed they will eventually be terminated to copper as a final metallic combination and therefore copper is used as the typical reference material. A useful purpose of this thermal EMF is the measurement of temperature using a thermocouple and microvolt meter.

A key feature of the foil resistor is its low thermal EMF design. The leads make close contact with the chip thereby maximizing heat transfer and minimizing temperature variations. The resistor element is designed to uniformly dissipate power without creating hot-spots and the lead material is compatible with the element material. Construction of the resistive element is such that the input and output junctions of the resistor are so close that temperature gradients between them is highly unlikely. These design factors result in a very low thermal EMF resistor. This is extremely important in DC circuits with low value resistors, usually below 1K ohm. Typical thermal EMF specifications are: 0.05 μV/°C, and 0.1 μV/°C Max, 1 μV/watt.

Application Information

Metal Foil resistors are used in high-tech electronic applications such as:

- computers
- medical equipment
- metrology
- high-end stereo equipment

- aerospace
- electronic precision scales
- laboratory
- test and measurement equipment

- telecommunications
- advanced instrumentations
- military
- structural stress / measurement

Hermetic sealing eliminates the ingress of both oxygen and moisture which degrade resistors over time.

A further hermetic enhancement in both short-term and long-term stability is achieved by oil filling. The oil also acts as a thermal conductor allowing the device to accept short periods of overload without degradation. These devices can be used as virtually secondary standards that can be carried in sets for daily or periodic calibration of factory measurement systems.

Foil Power resistors offer the best approach to low-value power and current sensing when a combination of accuracy, tight TCR, low thermal EMF, low voltage coefficient, Kelvin connection, and long-term stability (under power) are required.

Axial Leaded Foil resistors have the advantage of readily available auto-insertion equipment while the radial leaded devices may require additional tooling. Also, when converting from metal film to metal foil resistors, printed circuit boards may already be laid out for axial leaded devices.

Metal Foil Current Sensing resistors are direct replacements for certain wirewound devices but without the inductive characteristics of wirewounds. When the device is used in a four-lead Kelvin connection, there is no restriction on the lead lengths as would be the case if this was a two-terminal device. These devices are frequently used in oscillator circuits which require superior frequency stability.

Power and Current Sensing Metal Foil resistor typical applications are:

- noninductive design
- constant current power supplies
- character generation on CRT's
- electron beam controls
- forced balance electronic scales
- current sensing applications
- deflection amplifiers
- graphic display computers
- Wheatstone bridges

Power Plate resistors can dissipate large amounts (up to 25 watts) of power per square inch of resistive area when properly heat sinked. Current straps can be soldered to the copper terminal pads and voltage probes can be added for Kelvin connections. Thermal response is extremely fast due to the short thermal path through the ceramic to the heat sink. Thus, momentary overloads are possible without generating a dangerous internal hot-spot.

Power Plate low-ohm (0.25 ohm to 2 ohm typical) resistors are ideal for use as sense resistors in deflection amplifier circuits where low noise and high speed are essential in a high power, high stability application. For example, in a CRT radar sweep display the linearity and repeatability of the center point and the sweep can be kept at optimum as the related current through the sense resistor increases from zero to maximum current corresponding to center, zero, and edge of the display. Other typical applications include:

- CRT film recorders
- electron-beam integrated circuit mask generation
- graphics display computers
- flying spot scanners
- electron-beam recording equipment
- fire control radar display systems
- electron microscopes
- high speed video character display systems.
- heads-up display systems

Power Plate high-ohm (10 ohm to 5K ohm) resistors can be used as sense resistors in airborne fixed gyro navigation control computers. This resistor dissipates low power in normal flight, but when the aircraft negotiates a turn, the resistor experiences a surge of power which must be dissipated before the navigation system is again stable.

Matched sets of precision resistors provide an economical method of achieving matching in critical portions of the circuit without resorting to networks. Matching can be either by deviation at room temperature or by TCR tracking. Two or more discrete resistors are supplied as a set. Users should be cautioned that opening more than one set at a time has the potential for mixing sets and losing the factory match.

Metal Foil Discrete Surface-Mount Chip resistors offer an order-of-magnitude improvement in performance over other surface mount chip resistors. Caution: Soldering temperatures used during installation may cause resistance to shift up to 0.05%. Recommended installation processes are IR, vapor phase and convection reflow. Avoid the use of cleaning agents which could attack epoxy resins, which form part of the resistor construction.

A Molded Surface-Mount resistor in a rugged construction (welded internal lead connections) is capable of withstanding significant thermal cycling and mechanical stresses while maintaining its precision.

Surface-Mount Current Sensing Chip resistors have four tin/lead coated surface mount pads for a four-terminal connection where a current sensor is required. These small devices dissipate significant heat through the pads, so surface-mount users are encouraged to be generous with the pads and board traces. Some hybrid users may want to use conductive epoxy for the interconnections terminations.

Metal Foil Precision Chip resistors offer an improvement of accuracy and stability over other chip resistors in hybrid applications. The bonding of the gold wires to the chip has an effect on the overall resistance and on the temperature coefficient, according to the length of wire used. An available nomogram illustrates the change of resistance and TCR due to a length of gold wire added at wire bonding.

Metal Foil Networks hold their ratio tolerances under defined circumstances. Networks with tight ratio tolerances and controlled tracking extend the useful life of the equipment, whether expressed as mean-time-between-failure (MTBF) or service periods to recalibration or end-of-life cycle.

Networks with glass-to-metal seal headers offer good thermal dissipation and sharing of temperatures between resistors.

Networks in a ceramic dual-in-line package offer more pin availability and more chip capacity.

Networks in a ceramic flatpack offer the lowest profile, but take more printed circuit board space.

Networks in a ceramic leadless carrier are also available. However, when tight tolerance of low-ohm values is a consideration, fixturing and associated contact resistance must be taken into account.

Surface-mounting of a network in a ceramic package has the advantage of electrical isolation on the underside and in DIP form, a favorable pin arrangement when two networks are to be placed side by side and connected together.

Metal Foil Network applications include:

- differential amplifiers
- fuel metering systems
- spacecraft instrumentation
- resistive summing networks
- synchro input and summing networks
- binary ladders (A-D, D-A, current and voltage summing.)

- decoder networks
- heads-up displays
- naval weapons systems
- ratio and ratio-matching networks
- linear and linear-summing networks
- gain-defining resistors in digital voltmeters

- resolver networks
- fire control systems
- binary coded decimal ladder networks
- bridge networks
- ratio arms in bridge circuits
- quadrature bridge networks

Hermetic sealing of miniature networks enhances the already inherently stable environmental performance. The result is improved load-life stability and better performance during high temperature and moisture exposure.

Metal Foil Trimmers, when used as a variable resistor (rheostat), keep the resistance value constant independent of the changes in the ambient temperature.

When Metal Foil Trimmers are used in a voltage-divider mode, the two parts of the trimmer's element have different temperatures because of temperature gradients created by heat dissipated by the trimmer and by outside heat sources. The low TCR of metal foil minimizes the influence of these gradients on the resistances drifting apart; thereby, keeping the ratio essentially constant.

Metal Foil Trimmers are particularly useful in low resistance values (2 ohm to 1K ohm). They exhibit essentially infinite resolution, low TCR, no inductance, high stability with time, and no noise. They are used directly in series with the resistor to be adjusted, as opposed to parallel circuits and linearizing resistors used for trimming with other resistor technologies.

Metal Foil resistors can withstand considerable thermal shock including the accelerating effect of liquid nitrogen to hot oil testing. Many other constructions require a ramp-up and dwell to avoid destruction of internal connections.

Chapter 5

Wirewound Resistors

In wirewound resistors, three alloys are commonly used for the resistive element. They are nickel-chromium, copper-nickel, and gold-platinum. Nickel-chromiun is the most common due to its excellent temperature coefficient (less than ± 5 ppm/°C) and its availability in many different diameters. Copper-nickel is the next most popular, with a temperature coefficient of ± 20 ppm/°C. The gold-platinum alloy, that is actually a complex alloy of gold, platinum with small amounts of copper and silver has a high temperature coefficient of ± 650 ppm/°C, but has low resistance. That resistance is 85 ohms/cmf (cmf is a circular mil foot, a hypothetical quantity equivalent to one foot of wire that is 0.001 inches in diameter) while nickel-chromium has a resistivity of 800 ohms/cmf. The gold-platinum alloy can also withstand harsh environments.

The ceramic core of a wirewound resistor is either beryllium oxide, which has a high cooling capability, alumina (aluminum oxide) or steatite, which has the lowest thermal conductivity of the three materials but is low cost.

Wirewound resistors are used where large power dissipation is required and where AC performance is relatively unimportant. These devices are generally satisfactory for use at frequencies up to 20kHz. They are available with various insulating/moisture preventative coatings such as vitreous enamel, cement, molded phenolic, glass sleeves, or silicone.

Vitreous enamel units have excellent moisture-resistance properties and will not burn (although they may melt) under high overload conditions since they are made from a glass type material.

Silicone, which also has excellent moisture-resistance characteristics, is an organic material and is more flammable at lower overload conditions than vitreous enamel. It will also emit gases under overload conditions leaving deposits on electrical contacts.

Variable wirewound resistors have an element of continuous-length wire, wound linearly on a rectangular arc-shaped core, depending upon the style. The sliding contact traverses the element in a circular or straight line, again dependent upon style. The element is protected from detrimental environmental conditions by a housing or enclosure. The lead screw head is insulated from the the electrical portion of the resistor.

Fixed wirewound resistors have the resistance element which consists of a precisely measured (by ohmic value) length of resistance wire, wound on a bobbin or core (usually of ceramic). The resistance wire is an alloy metal without joints, welds, or bonds (except for splicing at midpoint of a bifilar winding and at end terminals). In order to minimize inductance, resistors are wound by one of the following methods: reverse pi-winding or bifilar winding. The element assembly is then protected by a coating or enclosure of moisture-resistant insulating material (usually inorganic vitreous enamel or a silicone) which completely covers the exterior of the resistance element including connections and terminations.

Power-type fixed wirewound, chassis-mounted resistors employ a construction of a measured length of resistance wire or ribbon (of a known ohmic value) wound in a precise manner (pitch, effective wire coverage, and wire diameter are controlled). These resistors must be wound either inductively or noninductively. Some use bifilar windings to reduce inductive effect. The continuous length of wire (wire required to be free of joints, bonds, and of uniform cross-section) is wound on a ceramic core or tube and attached to end terminations. End terminations are axial lug-type leads or radial tab styles. The finished resistor element and termination caps are sealed by a coating material. The coated element is then inserted in a finned aluminum alloy housing which completes the sealing of the element from detrimental environments, and provides a radiator and a heat sink for heat dissipation. Water-cooled housings are also available. These housings facilitate the transfer of heat away from the resistive element.

Power-type variable wirewound resistors have a resistance element of wire, wound on an insulating core and shaped in an arc. The wire and core are usually bonded to the base structure by a vitreous enamel. A contact arm bears uniformly on the resistance element when adjusted by a control shaft. Rotation is limited by stops, and electrical "off" positions are available. Some styles are classified as "unenclosed."

Hybrid potentiometers are wirewound units with a conductive-plastic track deposited along the contact path of the resistive element. That results in a device that has a better resolution and a longer life, by a factor of 10, over wirewound types. Compared to conductive-plastic units, hybrid devices have a higher power handling capability, due to the wirewound element. Like wirewound units however, they have stray capacitance at higher frequencies and have high contact resistance and marginal output smoothness when drawing current through the wiper contact.

For low resistance/high current applications, edgewound ribbon type power resistors are available. Designed for power handling up to 1000 watts (at currents up to 100 amps), these devices are made up of steel ribbons wound into a coil and supported by ceramic insulators. They are generally rated for normal operation with a temperature rise of 375°C. these types are used in power supply testing and in motor-breaking systems.

General characteristics of wirewound resistors:

- Fixed, wirewound, accurate resistors are physically the largest of all types for a given resistance and power rating, since they are very conservatively rated and are available in standard tolerances as low as ±0.1 percent.
- Because of the general method of construction (employing a plastic or ceramic bobbin), this type is subject to mechanical damage resulting from vibration, shock, and pressure.
- Used where high cost and size are not important and operational climate can be controlled.

- Application of voltages in excess of voltage rating may cause insulation breakdown in the thin coating of insulation between element coatings.
- Operation above 50kHz may produce inductive effects and also intrawinding capacitive effects.
- Resistance element is quite stable within specified temperature limits.
- Use of good soldering techniques is extremely important, since higher contact resistance may cause overall resistance shifts far outside of resistance tolerance on low value units.
- The presence of moisture may degrade coating or potting compounds.
- Fixed, wirewound, power-type resistors are generally not supplied in low tolerances, since most applications of this type do not require accurate resistance.
- The use of tapped power wirewound resistors is to be avoided, because insertion of taps weakens the resistor mechanically, and lowers the effective power rating.

Application Information

Fixed wirewound resistors are suited for use in electrical, electronic, communication, and associated equipment.

Fixed wirewound resistors are most often used in voltage divider circuits, as power supply bleeder resistors, or as series dropping resistors.

Variable devices can be used where voltage and current variations are expected. Applications include: motor speed controls; generator field controls; lamp dimming; heater and oven controls; potentiometer uses; and applications where variation of voltage or current is required (such as voltage-divider and "bleeder" circuits).

Precision variable types are used in servo systems requiring precise electrical and mechanical performance. They are used in computer, antenna, flight control, and bomb navigation systems for matching, balancing, and adjusting circuit variables.

Surface mount fixed wirewound resistors are compact in size, and are suitable for pick-and-place equipment. A fireproof inorganic construction using an inorganic potting compound provides high thermal conductivity and moisture resistance for use in aqueous board wash systems.

Fixed wirewound resistors are especially suited for use in:

- DC amplifiers
- Voltmeter multipliers
- Meters
- Laboratory test equipment.

Commercial, precision power, low value, fixed wirewound resistors are ideal for all types of current sensing applications including:

- Switching and linear power supplies
- Instruments
- Power amplifiers.

Commercial, power, fixed wirewound resistor applications include:

Kitchen appliances:

- Percolators
- Blenders
- Mixers
- Ranges
- Toasters
- Deep fryers.

Automotive devices:

- Horns
- Ignitions
- Windshield wipers
- Voltage regulators
- Instrument gauges.

Entertainment devices:

- Radios
- Televisions
- Computers
- Power supplies.

Industrial power (up to 225W) fixed wirewound resistors can dissipate large amounts of power in DC of low-frequency AC circuits. Applications include:

- Grid resistors
- Voltage dropping resistors
- High voltage bleeder resistors in power supplies
- Bias supply resistors
- Voltage divider networks
- Filament dropping resistors
- Load resistors
- Shunt resistors.

Noninductive applications are:

- Dummy antennas
- Terminating resistors
- Any function where low effective inductance is needed.

Fixed wirewound resistors are not designed for high-frequency circuits where their AC performance is of critical importance; however, provisions have been made in particular styles to minimize inductance.

Resistors are to be operated at rated ambient temperature. If resistors are to be operated at a greater ambient temperature, the resistor should be derated.

When using fixed wirewound resistors with low resistance values and a tolerance of 0.1 percent or less, the designer must consider the fact that the resistance of the leads and other wires connected to the resistor may exceed the tolerance. Where a resistor is used in a critical application that requires the initial tolerance to be 0.1 percent or less, it is also desirable to hold resistance changes within this tolerance during operation. Since the temperature characteristic can cause the resistance to change by more than 0.1 percent, the temperature rise in the resistor must be kept to a minimum if the resistor is expected to remain within the initial tolerance during use. It is to be noted that initial nominal resistance is measured at 25°C while full-load operating temperature is 125°C. Therefore, if this close tolerance of 0.1 percent or less is to be held, the power rating of the resistors should be reduced.

Because all of the electrical energy dissipated by a resistor is converted into heat energy, the temperature of the surrounding air becomes an influencing factor in the selection of a particular resistor for use in a specific application. After the desired resistance tolerance and the anticipated maximum ambient temperature have been determined, a safety factor of 2, applied to the wattage, is recommended in order to insure the selection of a resistor having an adequate wattage-dissipation potential, and one which will remain within specified tolerance limits.

When the resistor is to be used in a circuit where the surrounding temperature is higher than operation temperature or when the chassis area is restricted, the wattage must be reduced so as not to overload the resistor.

Where high voltages (250 volts and higher) are present between the resistor circuit and the grounded surface on which the resistor is mounted, or where resistance is so high that the insulation resistance to ground is an important factor, secondary insulation between the resistor and its mounting, or between mounting and ground, should be provided.

A solder with a low-melting temperature (300°C) should be used. Care must be exercised in soldering fixed wirewound resistors, particularly in the lower resistance values and tighter tolerances, since high contact resistance might cause resistance changes greater than the tolerance.

It is suggested that fixed wirewound resistors be mounted by restraining their bodies from movement when shock or high-frequency vibration forces are to be encountered. It should be noted that if clamps are used, certain electrical characteristics of the resistor will be altered. The heat-dissipating qualities of the resistor will be enhanced or retarded depending on whether the clamping material is a good or poor heat conductor. Under less severe vibration conditions, axial lead styles may be supported by their leads only. The lead lengths should be kept as short as possible, 1/4 inch or less preferred, but not longer than 5/8 inch. The longer the lead, the more likely that a mechanical failure will occur. For mounting of tab-terminal resistors, use of bracket assemblies is recommended.

Certain coating materials used in fabricating resistors may be subject to "outgassing" of volatile material when operated at high surface temperatures (200°C). This phenomenon should be taken into consideration during product design.

When stacking resistors, care should be taken to compensate for the added rise in temperature by derating the wattage rating accordingly.

When resistors are mounted in rows or banks, they should be so spaced that, taking into consideration the restricted ventilation and heat dissipation by nearby resistors, none of the resistors in the row or bank exceeds its maximum permissible continuous operating temperature. An appropriate combination of resistor spacing and resistor power rating must be chosen if this is to be assured. In view of the chassis heat dissipation principle of these resistors, particular care must be exercised in order that the chassis temperature rise does not damage nearby components.

Fixed wirewound resistors to be used in a product should be so chosen that, when mounted in the product, they will not be required to operate at a temperature in excess of their rating. This should be applicable under the most severe conditions as follows:

- In the maximum specified ambient temperature within the limited chassis area with all enclosures in place.
- Under conditions producing maximum temperature rise in each resistor.
- For a sufficient length of time to produce maximum temperature rise, or for the maximum specified time.
- With natural ventilation only. (This should permit the use of any special ventilating provisions included as a standard part of the product.)
- At high altitude or similar environment.

The use of resistive element wire size of less than 0.001 inch diameter is not recommended.

The maximum ambient temperature should not be exceeded.

Failures are considered to be opens, shorts, or radical departures from initial characteristics occurring in an unpredictable manner, and in too short a period of time to permit detection through normal preventive maintenance. Failure rate factors are based on "catastrophic failures" and will differ from the failure rates established based on a "parametric failure" of a percent change in resistance to be expected at hours of life tests at rated conditions.

Variable wirewound resistors are used primarily for noncritical, low frequency applications where characteristics of wirewound resistors are more desirable than those of composition resistors.

Variable wirewound types should not be used in frequency-sensitive RF circuits due to the introduction of inductive and capacitive effects.

A variable wirewound resistor's noise level is low compared to nonwirewound types. Peak noise is controlled at an initial value. However, after exposure to environmental tests (moisture, shock, vibration, etc.), a small degradation is allowed.

When a variable linear resistor is being used as a voltage divider, the output voltage through the wiper will not vary linearly if current is being drawn through it. This characteristic is usually called the "loading error." To reduce the loading error, the load resistance should be at least 10 to 100 times as great as the end-to-end potentiometer resistance.

Variable wirewound types are subject to noise because of stepping of the contact from wire to wire.

High humidity conditions may have an obscured or unexpected harmful effect on unenclosed types due to winding-to-winding shorts.

Good practice indicates the use of enclosed units to keep out as much dust and dirt as possible and to protect the mechanism from mechanical damage. The presence of oil from lubrication may cause dust or wear particles present to concentrate within the unit.

Since the resistance is variable, it is necessary to provide some method of preventing movement of the wiper arm, other than those movements required during operation. For resistors which are not in continuous use, the short locked shaft will limit the amount of motion due to shock, vibration, and accidental movement. Where it is absolutely necessary to have a long shaft, a coupled extension is preferred to one long integral shaft. Regardless of the type of shaft, the use of oversize control knobs which permit high rotational torque will generally result in damage to the integral stop. Use the smallest size knob to reduce torque.

Lead-screw actuated variable resistors can provide a high degree of accuracy in critical adjustments; however, the user should consider the effects of backlash in the lead-screw position versus wiper position. The resistance obtained at an initial setting may change slightly under conditions of vibration and shock as the wiper settles into a new position. The magnitude of this change is allowed to be as high as 1 percent when new, and can increase with age up to about 3 percent or the equivalent of one-half turn of the lead-screw. In extremely critical applications, it may be desirable to decrease the resistance value of the variable resistor, and add a suitable fixed resistance in series to obtain the same overall resistance, thus giving less critical adjustments but with a decrease in the adjustable range.

Adjustable wirewound resistors feature an adjustable resistor or voltage-divider. It can also be used to quickly obtain odd resistance values. Using a multi-tap resistor with one or more adjustable lugs can provide voltage-divider applications.

Care should be taken when moving adjustable lugs. The coatings protect the resistance wire from shifting and shorting to other turns during adjustment. However, the following three steps should always be taken whenever adjustments are made. Failure to follow these steps can result in damage to the resistor.

- Turn off current to avoid possible operator injury and damage to the unit.
- Loosen adjustable lug until it will slide completely free, without touching the exposed wire.
- When adjustment point has been selected, retighten lug only enough to assure a firm contact; do not tighten beyond this point.

The list below shows some of the options and combinations which can be achieved:

- **Construction:** Heat sink, silicone coated, epoxy or silicone molded (single or multi-element), hermetic seal (ceramic tube or metal can encapsulated), clip mounted or fireproof inorganic construction.
- **Leads:** Radial and axial type, special materials and dimensions, spaded, threaded, insulated, quick-disconnect eyelet, printed circuit, ferrule.
- **Matching:** By pairs or sets for T.C., tolerance or ratio.
- **Special Types:** Extended low or high resistance range, adjustable, low reactance, special wire alloys, very low or high T.C., high stability, special tolerances, tapped, water cooled, temperature sensitive, inductive.
- **Pre-conditioning:** Power aging, temperature cycling, temperature and power, short-time overload, thermal shock, X-ray, temperature aging.

- **Shunts:** Low value, four-terminal resistors for current sensing requirements.
- **Fuse Resistors:** Hybrid component designed to act as an ordinary resistor under normal circuit conditions, and as a fuse under fault conditions.

Chapter 6

Nonwirewound Resistors

Nonwirewound variable resistors have an element of metal, cermet type, or carbon film (depending upon characteristic) deposited or fused upon a ceramic or glass base. Depending upon style, the element is rectangular or shaped in an arc and the sliding contact maintains continuous contact when traversing the element in a straight line or circular motion. The element is protected from detrimental environmental conditions by a housing or enclosure. The lead-screw head is insulated from the electrical portion of the resistor.

Precision nonwirewound variable resistors have a resistance element usually consisting of carbon, cermet, or conductive plastic deposited on a plastic insulating base. Conductive plastic is a generic term covering a broad category of materials and manufacturing methods. It includes the "bulk" type compression molded materials and the oven cured thick films (screened, sprayed, dip coated, or roll coated). All of these conductive plastic materials invariably utilize carbon as the resistive material together with a resin binder and an inert filler. The moving contact is insulated from the operating shaft and maintains continuous electrical travel throughout the entire mechanical travel. The electrical output (in terms of percent of applied voltage) is linear with respect to the angular position of the operating shaft. The element and contact arm are enclosed in an environmentally resistant housing.

Cermet devices have a resistive element made by combining very fine particles of ceramic or glass with precious metals. Cermet devices are very stable under humid conditions and have low temperature coefficients of ± 100 ppm/°C. Conductive-plastic or hot-molded carbon potentiometers, for example, have an average temperature coefficient of ± 1000 ppm/°C.

In variable resistors, however, the cermet element is abrasive and long periods of rotational cycling will wear out the wiper long before similar use would wear out the wiper in resistive-film or conductive-plastic units.

Cermet potentiometers are available in low resistance values, which makes them useful in many audio applications.

Cermet is also the thick film used in resistor networks and in chip resistors.

Conductive plastic potentiometers have a resistive element consisting of a blend of resin (epoxy, polyester, phenolics, or polyamides) and a carbon powder applied to a plastic or ceramic substrate. The plastic substrate results in a better temperature coefficient due to greater compatibility between the ink and the substrate. Those devices have a long rotational life and excellent contact resistance variation, or low noise. End resistance is low, two ohms maximum.

Conductive plastic units are suitable for use in applications that require a consistent temperature coefficient over a limited temperature range, such as -25°C to +75°C. A temperature coefficient of ± 1000 ppm/°C is average. Temperature coefficient values of -200 ppm/°C may be attained by special processing of the carbon material or by incorporating metal powders, or flakes, into the element. Nickel, silver, and copper are frequently used in low-resistance devices. Conductive-plastic elements, like carbon units, vary in resistance when exposed to humid conditions.

Application Information

Use variable nonwirewound resistors for matching, balancing, and adjusting circuit variables in computers, telemetering equipment, and other critical applications.

Precision variable nonwirewound resistors are required in servo-mounting applications for precise electrical and mechanical output and performance. They can be used in computer, antenna, flight control, and bomb navigation systems, etc.

Some variable nonwirewound resistors are suitable for rheostat or potentiometer applications where high precision is not required, and are capable of withstanding acceleration, shock, high-frequency vibration, and elevated operating temperature at rated load. They are most useful in circuitry where high resistance values and low power dissipation are encountered in controlling volume, bias, tone, voltage output, and pulse width. These resistors have a film resistance element that is rectangular or shaped in an arc, and a contact bearing uniformly thereon, so that a change of resistance is produced between the terminal of the contact and the terminal at either end of the resistance element when the operating shaft is turned. In some styles, the construction of the element is metal-ceramic film fused onto a ceramic substrate. The element is then contained in an enclosure which provides for environmental and mechanical protection.

Wattage applied during specified applications should be taken into consideration. When considering these resistors for potentiometer applications, it is necessary to bear in mind the fact that the load current as well as the "bleeder" current will be flowing through a part of the resistor and will contribute to the heating effect.

Consideration should be given to temperature rise and ambient temperature of resistors under operation in order to allow for the change in resistance due to resistance-temperature characteristic. This characteristic is measured between the two end terminals. Whenever the resistance-temperature characteristic is critical, variation due to the resistance of the movable contact should be considered.

When a resistor is to be used where the surrounding temperature is higher than the applicable operating temperature, a correction factor must be applied to the wattage rating so as not to overload the resistor. This correction factor may be taken from the derating curve.

After the anticipated maximum ambient temperature has been determined, a safety factor of 2 should be applied to the wattage is recommended in order to insure the selection of a resistor style having an adequate wattage with optimum performance.

The maximum continuous working voltage specified should in no case be exceeded, regardless of the theoretical calculated rated voltage.

Where voltages higher than 250 volts rms are present between the resistor circuit and grounded surface on which the resistor is mounted, or where the DC resistance is as high as the insulation resistance to ground, secondary insulation to withstand the conditions should be provided between the resistor and mounting or between the mounting and ground.

Resistors should not be mounted by their flexible-wire leads. Mounting hardware should be used. Printed-circuit types are frequently terminal board mounted, although brackets may be necessary for a high-shock and vibration environment.

When stacking resistors, care should be taken to compensate for the added rise in temperature by derating the wattage rating accordingly.

The noise level for nonwirewound resistors is quite low compared to composition or wirewound variable resistors.

Linear resistance taper is one having a constant change of resistance with angular rotation.

Nonlinear resistance taper has a variation or lack of constancy in the change of resistance with angular rotation.

Failures are considered to be opens, shorts, or radical departures form initial characteristics occurring in an unpredictable manner, and in too short a period of time to permit detection through normal preventive maintenance.

Chapter 7

Shunts, Current Shunts, and Current Sensors

Generally speaking, there exists considerable overlapping in the usage of the terms "Shunts," "Current Shunts," and "Current Sensors." Without totally discouraging the interchangeable usage of these terms, there are some subtle differences which are worth noting and which may suggest a preference for one or another term as it relates to a particular application.

SHUNT - "A resistive device employed to divert most of the current in an electric circuit." The earliest shunts were meter shunts used as external accessories to ammeters allowing one meter to be used for a variety of current levels depending upon which shunt was chosen. These were often massive four-terminal devices with the smaller potential terminals connected to the meter and the larger current terminals connected to the circuit under test. Present day ammeters are more likely to be specific to a particular current range - one meter, one internal shunt. These internal shunts are resistors with current connections to the external terminals of the ammeter and voltage connections made internally to the meter movement.

In addition to these measurement shunts, power shunts are used for electric motor starting, braking, and speed control. Loading, neutral grounding, preheating, and capacitor unloading are all applications in which a resistor is called upon to shunt large amounts of current.

A shunt may be a two-terminal or four-terminal resistor of sub-ohmic resistance value and high current capacity.

CURRENT SHUNT - "A four-terminal shunt." A current shunt shall be capable of current sensing as well as conducting significant current.

CURRENT SENSOR - "A resistive device employed to sense levels of current." Current sensors are used in applications where the emphasis is on accuracy and repeatability under all conditions and less on high current capability. Applications such as force-balanced scales, E beam deflection systems, switching power supplies, etc. all rely on current sensors to feed back and control the current. A current sensor is a four-terminal resistor of high accuracy and stability and usually of low resistance value.

Kelvin Connection

Kelvin or four-terminal connection is required in these low ohmic value products to measure a precise voltage drop across the resistive element. In these applications, the contact resistance and the lead resistance may be greater than that of the element resistance itself so lead connection errors and lead temperature coefficient of resistance errors can be significant if only two-terminal connection is employed. Figure 6.1 shows a resistor in series with an inductor in parallel with a capacitor including the resistance and inductance of the leads. This is the equivalent circuit of a two-terminal resistor.

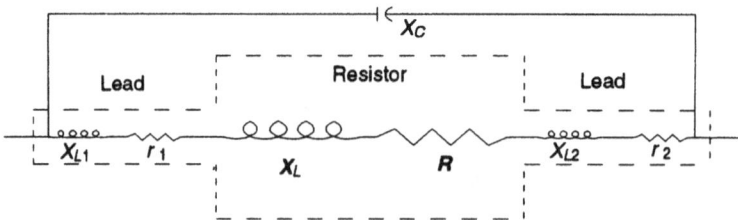

Figure 6.1 The Equivalent Circuit

By ignoring the inductance and capacitance for now, the two-terminal resistor is now shown in Figure 6.2. If $r_1 = r_2 = r$ then the total resistance $R_T = R + 2r$. The lead resistance r is uncertain because there is no user-assured connection to the lead. Thus if we allow r to be significant compared to R, small inaccuracies in lead connections become large inaccuracies in reading. Furthermore, since the lead material is likely to be copper [with a resistance change with temperature (TCR) of +3900 ppm/°C] and the shunt might be manganin [with a TCR of -20ppm/°C], then r is very large compared to R and the device is useless in a temperature variable application.

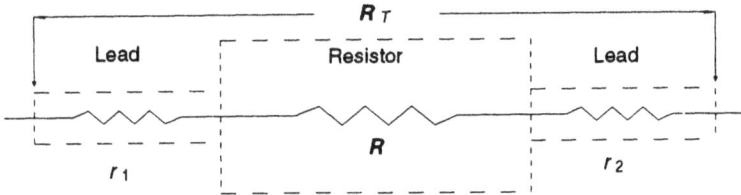

Figure 6.2 The Simplified Circuit

Lord Kelvin in the 19th century developed, among other things, the four-terminal method of measurement which eliminated both the uncertainty of lead resistance and lead response to temperature. Figure 6.3 is the Kelvin solution. If the voltage measuring system used here is of a high impedance, then r_5 approaches infinity and the measurement current I_m approaches zero. With zero I_m there is zero IR drop through r_3 and r_4 and therefore it does not matter whether the contact resistance is large or small. It also does not matter if they have a high TCR. Similarly, the TCR of the current leads is no longer important because the voltage connections are fixed inside the

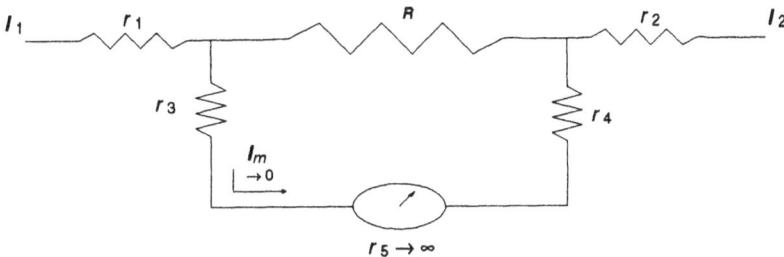

Figure 6.3 Kelvin Connection

lead resistance and in this way, the resistance, contact resistance, and/or lead TCR are thus eliminated. Kelvin connection to a four-terminal resistor is essential for precise current sensing.

Measurement of Resistance at Low Values

The ability to measure low values to tight tolerances is a concern to both the manufacturer and the user and in many cases coordination of reading ability by exchange of serialized units with recorded readings becomes necessary. The problem is compounded in the cases of high current shunts where the self-heating will cause the "in-service" resistance value to be different than that obtained with low current level measuring equipment, therefore, the measurement conditions must be defined (agreed upon) at the time of specification preparation, i.e., resistance value as determined by specified current and measured IR drop following a period of stabilization.

Measurement equipment is available from a number of sources with varying stated accuracies. Digital multimeters with 5 1/2 to 8 1/2 digits may have a 1 ohm full scale range; and if the stated accuracy is sufficient, these devices are suitable for direct reading of resistance down to 0.001 ohms when equipped with Kelvin connections. If indirect readings (calculated from current and voltage readings) are acceptable, then the digital multi-meter can be switched from the ohms to the voltage function and with a constant current power supply across the known and unknown resistors in series, IR drops across the potential leads can be measured and compared. this method permits measurement at rated current which is not available with the "ohms" function of multimeters and/or most low ohm bridges. Thus, measurement with the higher rated current assures correct results at anticipated operating conditions. One useful variation of this scheme is to fixture and read many resistors at one time by passing the constant current through all resistors in series and switching the voltage probes from one resistor to the next. Large quantities are thus measured quickly. It should be noted that manufacturers will generally prefer a resistance specified at room temperature and not have to deal with the stabilization time necessary for rated current readings. If rated current readings are required, a test charge may be imposed.

Application Information

Shunts

Shunts are suitable for precise current sensing and should be considered when a power requirement exists in addition to the precision and stability of a current sensor.

Measurement shunts are used in ammeters.

Power shunts are used for electric motor starting, braking, and speed control, also loading, neutral grounding, preheating, and capacitor unloading applications.

Shunts that are low power (1-10W) are purposely under-rated to accommodate the tight tolerance and TCR expected of them. Good practice calls for derating with tolerance to hold the time drift to a level consistent with the initial tolerance.

Design factors for shunts:

- **Power rating required** - How much power will be dissipated?
- **Disposal of heat generated** - What services are available to remove heat? Forced air?
- **Mounting surface and hardware** - What shock and vibration conditions dictate standard or special mounting?
- **Resistance value and tolerance** - The tolerance must be consistent with the ability to measure. With specific electrical gauge points identified, two-terminal devices can be toleranced to discriminate 0.001 ohms. Four-terminal devices can be toleranced to discriminate down to 0.0001 ohms. Thus, a two-terminal resistor of 0.1 ohms can be given a tolerance no tighter than ±1.0% (1.0% of 0.1 ohms = 0.001 ohms) while a four-terminal resistor can be given a tolerance as tight as ±0.1% (0.1% of 0.1 ohms = 0.0001 ohms).

To fully describe a particular shunt, the following characteristics may require identification:

Mechanical (as applicable)

- All external dimensions.
- Body material and finish.
- Terminal material and finish.
- Mounting method and hardware.
- Maximum internal hot-spot temperature.
- Maximum case temperature and test location (related to hot-spot temperature).
- External conductors, termination, and method of attachment.
- External heat sink capacity and mounting method.
- Cooling water connection with specified inlet temperature, flow rate, and pressure.
- Part marking.

Electrical (as applicable)

- Resistance value and tolerance (for four-terminal devices, also the maximum resistance value between the current terminals).
- Maximum current rating.
- Maximum continuous duty wattage rating based on specified heat sinking method.
- Maximum pulse power under specified conditions.
- Maximum temperature coefficient of resistance and temperature range and reference.
- Maximum reactance.
- Maximum dielectric withstanding voltage to mounting hardware.

Sensors

Current sensor applications include circuits that feed back and control the current in switching power supplies and E-beam deflection systems.

Some current sensors are plastic encapsulated while others are in metal housings and still others are hermetically sealed. Metal enclosures provide a more rugged mounting and more thermal conduction. Hermetic enclosures improve long-term stability.

Radial leaded current sensors take up less board space for a given size. Axial/radial lead configurations have the advantage of short potential connection leads to the PC board thus minimizing thermal EMF generation.

Exposed element shunts/current sensors are not the most accurate devices prior to installation but once installed, can be extremely stable and with microprocessor correction for initial accuracy, they represent a very cost effective approach to current sensing.

Shunts/Current sensor chips for hybrid packaging and surface mounting practitioners will find these leadless devices to be a solution to current sensing in a small space.

Shunts/Current sensor low-ohm networks represent an approach to switchable output when the voltage to current level must be variable.

Design factors for sensors:

- **Current Rating** - How much current is the sensor expected to handle? Continuously? Pulsed? Distortion restricted?
- **Sensitivity** - How much output signal is required per unit of current? (How much μV or mV per ampere?) This sets the resistance value and the power rating.
- **Resistance Value** - The required sensor resistance is equal to the current rating times the sensitivity divided by the current rating or numerical equivalence to the sensitivity, but in ohms. Example: A 2A sensor with 100 mV/A sensitivity will have a full scale output of 0.2 volts. $R = E/I = 0.2/2 = 0.1$ ohms.

- **Power Rating** - The required power rating may be greater than the product $E_{max} I_{max}$ if the sensitivity must be the same following self-heating. A positive TCR will cause the resistance value to increase along with the sensitivity. This effect can be minimized by overspecifying the power rating or moving to a lower TCR element.

To fully describe a particular current sensor, the following characteristics must be identified:

Mechanical (as applicable)

- All external dimensions.
- Body material and finish.
- Lead material and finish.
- Identification of current and potential leads.
- Part marking.

Electrical (as applicable)

- Resistance value and tolerance - Alternate specification: Voltage output and tolerance for a given current through: e.g. 50 mV $\pm 0.1\%$ per Amp.
- Maximum temperature coefficient of resistance.
- Maximum wattage rating.
- Maximum reactance.
- Maximum dielectric withstanding voltage.
- Maximum current noise.
- Maximum thermal EMF at the terminals.

Sub-ohm standards are offered to assist in the measurement of the shunts and current sensors. Low ohm measurements are subject to errors from various causes. These secondary standards are not adjustable but with the certificate of accuracy can be used with a comparator bridge to measure sub-ohmic values quite accurately.

Chapter 8

NTC Thermistors

The term "Thermistor" is a contraction of the words "Thermally Sensitive Resistor." This phrase accurately describes the basic function of the component, that is, to exhibit a predictable change in electrical resistance as a function of any change in absolute body temperature.

The first thermistors were developed simultaneously in the United States and Holland during World War II. Those first thermistors exhibited the high negative temperature coefficient of electrical resistance over an extended temperature range just as thermistors do today. But those early devices were not stable or reproducible. Due to improvements in raw materials and better manufacturing techniques, contemporary thermistors are highly stable, reproducible, and precise. Now, the term thermistor is a generic name for any device made from a semiconducting material that possesses an electrical conductivity which is highly sensitive to temperature.

No other electrical component exists in such a wide variety of physical shapes and sizes as the thermistor. Each mechanical configuration can possess a range of up to 20 different electrical properties. New combinations of electrical and physical properties are constantly being tested and developed by research and development teams. While other technologies may have arisen, thermistors continue to outperform their competition and to remain absolutely critical to production in many different kinds of industries. Suitable replacements, alternatives or substitutes for thermistors are just not available. A variety from sub-pinhead sized beads through disc, rod and flat film shaped to metal-sheathed probes and plastic encapsulated assemblies offers an advantage for any application.

The NTC (Negative Temperature Coefficient) thermistor is a relatively simple, two-terminal semiconductor made from sintered mixtures of metallic oxides such as manganese, cobalt, and nickel. Unlike the resistance/temperature characteristic of most metals, the NTC thermistor decreases nonlinearly in resistance as a result of an increase in body temperature.

Changes in the body temperature of the NTC can be brought about:

- **Externally** - by a change in the temperature of its surroundings (ambient heating).
- **Internally** - by heat resulting from wattage developed by the passage of current through the device (self-heating), or by a combination of these effects.

Thermistor types are the glass bead, glass encapsulated, glass probe, or small disc thermistors.

Thermistor probes, already installed in a variety of industrial and commercial applications, are appealing to a growing number of designers for sensing the temperature of gases and liquids as well as surfaces.

As a general rule, the lowest cost thermistor probes utilize a disc or glass encapsulated thermistor in an aluminum or molded epoxy housing. There is virtually no end to the design configuration of probe encasements. You can increase the maximum operating temperature by using insulated leads and/or high-temperature solder on all internal connections. A high operating temperature thermistor uses a stainless steel housing with a glass bead or a glass probe thermistor with welded internal connections of solid nickel and insulated wire.

Thermistor Terminology

A "Zero-Power" or "No Load" Resistance/Temperature curve is an exponential-type curve. It is a plot of measured thermistor resistance as a function of ambient temperature only; i.e., applied wattage held to a negligible amount in order to minimize the effects of self-heating. This is accomplished by submerging the thermistor in a temperature-controlled oil bath (held to $\pm 0.005°C$) and measuring the resistance with a low-power bridge.

There are three parameters that are generally used throughout the thermistor industry to describe the basic characteristics of any NTC:

- **Zero-Power Resistance @ 25°C** - the resistance of a thermistor at 25°C with zero electrical power dissipation.
- **Resistance Ratio** - the ratio of measured resistance at any two reference temperature points.
- **Temperature Coefficient of Resistance** - a measure of the slope of the R/T curve at any point with respect to the resistance at that point. The temperature coefficient is expressed in ohms/ohm/°C or, more commonly, %/°C.

When designating a particular NTC material system, the temperature coefficient of resistance at 25°C is often used.

Resistance vs. Temperature Characteristic

The amount of change per degree Celsius (°C) is defined by either the BETA VALUE (material constant), or the ALPHA COEFFICIENT (resistance temperature coefficient). Beta is expressed in degrees Kelvin (K). Alpha value is usually expressed in percent per degree Celsius (% /°C). The larger the Alpha or Beta Value, the greater the change in resistance per degree Celsius, or the "steeper the curve."

Within the thermistor industry, a thermistor material system is usually identified by specifying the Alpha coefficient, Beta Value, or the ratio between the resistance at two specified temperatures. Typical examples: R0/R50 or R25/R125.

Electrical Resistivity

Electrical Resistivity (ohm-cm) is a property characteristic of different materials. It is equal to the resistance to current flow of a centimeter cube of a particular material, when the current is applied to two parallel faces.

When comparing different thermistor materials. it should be noted that the material with the larger Alpha or Beta value will generally have the larger resistivity.

Material resistivity is an important consideration when choosing the proper thermistor for an application. The material must be chosen such that a thermistor chip of a specified resistance value will not be too large or too small for a particular application. Thermistor materials are available with a variety of resistivity values. The resistance of an NTC thermistor is determined by the material resistivity and physical dimensions. Required resistance value is usually specified at 25°C.

A chip (0.040" square by 0.015" thick) fabricated from one material may have a resistance @ 25°C of 10,000 ohms nominal, while a chip fabricated from a different material with the same dimensions could have a resistance of about 100,000 ohms @ 25°C.

Current vs. Voltage Characteristic

At low currents, the power dissipated by a thermistor is quite small, and the thermistor behaves like a fixed resistor. When plotted, the voltage drop vs. current will be linear.

Once the current reaches a certain level, the thermistor can no longer dissipate all of the power being generated, and self-heating occurs. Starting with the maximum power which can be dissipated without self-heating, if the current is further increased, voltage drop will begin to decrease with current. If the current is allowed to increase unchecked (without a current-limiting resistor), the thermistor resistance will decrease, and the device will overheat to the point of failure.

The self-heating properties of a thermistor are utilized in many applications, including flow measurement. For most temperature measurement applications, to reduce measurement error, it is advantageous to keep self-heating to a minimum.

Current vs. Time Characteristic

When a thermistor is subjected to a step change in power, there is a time lag before a steady state current condition is reached. During this time lag, the current will rise, reaching an equilibrium after time. The current time characteristic is influenced by thermistor dissipation constant, geometry, circuit design, mounting method, and environment.

Dissipation Constant

The dissipation constant of a thermistor, expressed in milliwatts per degree Celsius (mW/°C), is defined as the power required to raise the thermistor body by one degree Celsius. The dissipation constant of a thermistor is influenced by geometry (size and shape factors), mounting method, thermistor material properties, thermal conductivity of thermistor environment, and other factors.

Thermal Time Constant

The thermal time constant of a thermistor characterizes a thermistor's response to step changes in temperature. It is especially useful when comparing the relative thermal response properties of different thermistor assembly designs. The thermal time constant is defined as the time required for a thermistor to change 63.2 percent of the total difference between its initial and final body temperature when subjected to a step function change in temperature.

The industry standard method for determining the thermal time constant is the self-heating method. This is done by mounting the thermistor in a constant-ambient, standard-heat-sink, zero-air- flow environment, and then applying enough power to raise the body temperature to a predetermined temperature. At that point, power is removed and the time constant is measured as the device cools to 63.2 percent of the temperature differential between the final and ambient temperature.

The self-heat method essentially defines a recovery time for the NTC. In applications such as temperature measurement, the catalog-listed value for the thermal time constant gives an indication of the thermistor's ability to change in resistance, due to a step-function change in ambient temperature. However, in self-heating applications, the thermal time constant is not nearly as significant as the value for the heat capacity of the device (H).

Thermistors intrinsically exhibit a predictable, logarithmic shift in resistance which is a function of time. State-of-the-art manufacturing techniques are designed to produce a finished product which is highly stable and exhibits minimal drift. Stringent process and raw material control is essential. These controls assure that each thermistor manufactured will have the high stability required for critical applications.

While no thermistor curve standards have been established by convention, material systems which have particular resistance vs. temperature characteristics have become de facto standards. Therefore, to insure accuracy and reproducibility, it is essential that the resistance vs. temperature characteristics of any thermistor be consistent and reproducible. Matched wafer thermistors are designed to closely track an established resistance vs. temperature curve. The other thermistor elements manufactured (point matched wafer thermistors, glass-coated bead thermistors, and bead-in-glass probes) are designed to curve track per typical industry standards.

The tolerance on resistance at any temperature is the sum of the closest tolerance at any specified temperature, and the additional tolerance spread due to deviation in resistance ratio.

In any application where the NTC thermistor is to be used to measure temperature, it is usually more appropriate to determine temperature tolerance rather than resistance tolerances. It is easily done by using the temperature coefficient of resistance.

Application Information

The range of NTC thermistor applications is astoundingly broad. From the de-icing devices on the wings of an airplane to the highly sensitive temperature monitoring system in a newborn baby's incubator, NTC thermistors are being used often by engineers and industrial designers because of the thermistor's ability to measure temperature accurately and inexpensively.

Industrial controls, medical sensors, automotive components, air conditioning units, digital thermometers, aerospace and military applications, HVAC, custom assemblies, probes, wafers, interchangeable components... super stable NTC thermistors are found in a wide variety of machinery, medical devices, equipment, appliances, and computers. While other technologies have arisen lately, the thermistor continues to prove itself to be a mainstay of many industries especially in the medical field. Designers rely on thermistors because of their stability, price, and the ease in which they can be incorporated into applications.

A thermistor's high sensitivity to temperature eliminates the need for expensive amplification schemes or four-wire measurements.

The main disadvantages of thermistors are nonlinearity and low power handling capability. These factors presented a challenge to the circuit designer in the past, but contemporary microprocessor designs have greatly simplified circuit design.

The measurement and control of temperature is a fundamental requirement of a variety of industries, from chemical processing to medicine. Thermistors are the sensor of choice for many applications. NTC thermistors offer an inexpensive, highly sensitive alternative to thermocouples, RTD's, and other types of sensors. Unlike other sensors, inexpensive interchangeable thermistors are readily available. Thermistors are highly sensitive to relatively small changes in temperature. Figure 7.1 on the following page compares the properties of NTC thermistors to other types of temperature sensors.

SENSOR TYPE	NTC THERMISTOR	RTD	THERMO-COUPLE	I.C. SENSOR
PARA-METER	Resistance vs. Temperature	Resistance vs. Temperature	Voltage vs. Temperature	Voltage or Current vs. Temperature
A D V A N T A G E S	• Large Change in Resistance vs. Temperature • Fast Time Response • High Resistance Eliminates the Need for 4-Wire Measurement • Small Size • Inexpensive • High Stability • Interchangeable to Tight Tolerance	• Linear • High Stability • Wide Operating Temperature Range • Interchangeable Over Wide Temperature Range	• Wide Operating Temperature Range • Simple • Inexpensive • Rugged • No External Power Supply Required	• Linear • High Output vs. Temperature • Inexpensive
D I S A D V A N T A G E S	• Nonlinear • Operating Temperature Limited to Approximately -60 to +300 °C • Interchangeable Over Relatively Narrow Temperature Ranges • Current Source Required	• Small Change in Resistance vs. Temperature • Relatively Slow Time Response • Low Resistance Requires 3- or 4-Wire Measurement • Sensitive to Shock and Vibration • Current Source Required • Expensive	• Nonlinear • Relatively Low Stability • Low Sensitivity • Low Voltage Output Can Be Affected By RFI and EMI • Reference Junction Compensation Required	• Limited Operating Temperature Range • Subject to Self-Heating • Limited Configurations

Figure 7.1 Comparison of Common Temp. Sensing Technologies

Protection against extremely high peak inrush current, especially in AC/DC switching power supplies, is now available with inrush current limiters. These special NTC thermistors effectively control surge currents because the thermal time constant of the current limiter is longer than the electrical time constant (RC) of the thermistor and the capacitor.

Glass encapsulated NTC thermistors are tiny pellets of thermistor material hermetically sealed in glass bodies with a controlled amount of selected atmosphere. The hermetic seal makes these parts extremely stable in applications up to 200°C.

Glass bead and glass probe units are very small glass-covered thermistor discs with the leads molded directly into the disc. The glass seal protects the thermistor from hostile environments providing excellent stability up to 260°C. These thermistors are used mostly for temperature measurement and control applications, and are usually mounted or potted inside of a probe or assembly.

Precision thermistors are widely used in OEM applications because they offer close-tolerance temperature measurement without requiring individual calibration.

Certain high-temperature thermistors are designed specifically for temperature measurement and control applications ranging from 200°C to 1000°C. These are well suited for applications such as self-cleaning ovens and are especially useful for monitoring exhaust gas temperature in automotive and aircraft applications.

Cryogenic thermistors are extremely useful for liquid level detection in various cryogenic liquids such as oxygen or nitrogen. In this application, the thermistor is slightly self-heated by passing a small current through the unit. The heat generated in the unit is more easily dissipated when the thermistor is immersed in cryogenic fluid than when the fluid level falls below the thermistor. The resulting change in thermistor temperature is easily detected by the change in resistance.

NTC chip thermistors are available for applications such as SMD, hybrids, temperature sensing, time delay, amplitude control, and temperature compensation.

Thermistor applications can generally be grouped into one of two categories:

- Zero Power Applications
- Self-Heating Applications.

Zero Power Applications

Temperature Measurement - The high sensitivity of a NTC thermistor makes it an ideal candidate for low-cost temperature measurement. One of the simplest thermometer circuits consists of a NTC thermistor as one leg of a Wheatstone bridge. Substituting a second thermistor makes the circuit twice as sensitive, permitting the use of a lower sensitivity meter.

Because the resistance vs. temperature characteristic of the NTC thermistor is nonlinear, it is often advantageous to linearize the curve. A simple voltage divider tends to linearize the output voltage as a function of temperature, while a single parallel resistor linearizes the resistance vs. temperature curve. In each case, the maximum linearity error is a function of the length of the temperature span. For spans of less than 50°C, a single resistor can linearize to better than ±0.5°C accuracy.

Decreasing the total linearity error can be accomplished by using more than one thermistor in the network. If the temperature span is relatively short, it is also possible to improve accuracy by using a single thermistor in a series/parallel resistor network.

Contemporary microprocessors have made it possible to design high accuracy temperature measurement circuits, while keeping circuit elements to a minimum. The nonlinear output of a thermometer can be converted directly to a temperature output.

There are a variety of off-the-shelf devices available for utilizing thermistors for measurement and control. A thermistor input process monitor/controller accepts standard thermistor inputs, and can be easily used in a variety of temperature indication and control applications.

Temperature Control and Temperature Alarm - A NTC thermistor can be utilized for temperature control and alarm applications using a minimum of electronic circuitry. A simple temperature controller is obtained by placing a thermistor in series with a relay coil and potentiometer. The potentiometer will control the switching temperature. For a more sensitive controller, a Wheatstone Bridge circuit is one of the simplest methods available to the designer. If the indicating meter is replaced with a relay, or other suitable triggering device, a temperature control or alarm is created. The bridge is balanced so that at steady state conditions, current flow across the relay is negligible. When the temperature of the thermistor deviates, the bridge is imbalanced, causing current flow, and relay actuation.

When used in conjuction with an amplifier, the thermistor provides a low-cost means of achieving highly reliable temperature control. The system can be as simple as the on/off control of a transistor driving a relay or as sophisticated as a closed loop proportional controller. The thermistor's main asset in temperature control applications is its high degree of sensitivity. Using thermistors, temperatures have been controlled to better than $0.001°C$.

Temperature Compensation - Because most conductors have a positive temperature coefficient of resistance, the NTC thermistor is often used to compensate such items as copper coils in D'Arsonval meters, etc. The method is to match the temperature coefficient of the thermistor to the compensated material. Where a copper meter coil would change 50% in resistance over a commonly used temperature range, a thermistor shunted by a resistor in series with the unit allows the total impedance of a circuit to be held uniform over the entire operating range. Due to the high temperature coefficient of the thermistor as opposed to the low temperature coefficient of the copper, full compensation can be achieved by using a thermistor-resistor network. This network adds less than 15% to the total impedance of the circuit. Compensation of transistor amplifiers, crystal oscillators, etc., can be achieved by using similar methods.

Self-Heating Applications

If a NTC thermistor is held at steady state conditions (constant current, constant physical surroundings, etc.), the thermal dissipation will be constant. If some physical parameter is changed, the thermal dissipation and the voltage drop across the thermistor will change. A correlation can be made between the change in voltage drop and the changing physical characteristic.

Fluid Flow and Level Applications - When a NTC thermistor is moved from still air to still fluid, for example, the dissipation constant of the thermistor will increase, and the voltage drop will be much higher. If the fluid is forced to flow, the voltage drop will increase further. This concept is the basis for thermistor fluid flow and level sensing.

The NTC thermistor can be used to sense the absence or presence of a liquid by taking advantage of its difference in dissipation constant between a liquid and a gas. In general, a self-heating thermistor in a liquid can dissipate roughly four to six times as much power as it can in air. Likewise, a self-heated or indirectly-heated NTC device can shed much more heat in air flow than it can in still air. A liquid-level/air-flow circuit usually looks for the resulting difference in voltage drop across the thermistor to light a lamp, turn on a buzzer, or as input to a comparator. A series resistor with the thermistor is to limit the amount of current available to heat the thermistor.

A good liquid-level/air-flow design should ensure that the system functions at the worst-case operating point. In the liquid-level application, the thermistor must be able to dissipate more power when submersed in the hottest liquid than it can when moved into the coldest air. A second thermistor can be added to compensate for the ambient. A similar situation exists in air-flow applications where the design should ensure that the thermistor can shed more heat in the hottest flowing air than it can in the coldest still-air ambient.

Time Delay/Surge Suppression - A NTC thermistor subjected to a step change in power will experience a time lag before reaching a steady state condition. This time lag can be utilized for surge suppression purposes. NTC thermistor resistance will initially be high, and current flow to the circuit will be limited. As the thermistor self-heats, resistance decreases, resulting in increased current flow.

By placing a thermistor in series with a relay, a potentiometer, and a battery, a simple time delay circuit is obtained. A relatively high potential is applied to the circuit. The thermistor begins to "self-heat," lowering its resistance and allowing more current to flow. The increased current further heats the thermistor, allowing still more current to flow, which in turn actuates the relay. The time can be controlled by adjusting the potentiometer.

A NTC thermistor in series with a light bulb can operate as a shock absorber for the light bulb. Because of the low cold resistance of the lamp filament, initial surge currents at turn-on are extremely high, causing degradation of the filament. With the NTC in the circuit, a high percentage of the available voltage is dropped across the NTC. As the thermistor heats, its resistance decreases quickly and more voltage is applied to the light bulb. This "soft start" technique can significantly increase the expected life of an incandescent lamp.

A similar application is found in switching power supplies where the thermistor is used to limit the excessive inrush current at power-on. In this case, the NTC does not protect the power supply as much as the series elements (relays, contacts, switches, and fuses) that are upstream from the power supply. The problem of current surges in switch-mode power supplies is caused by the large filter capacitors used to smooth the ripple in the rectified 60Hz current prior to being chopped at a high frequency. The NTC thermistor is ideally suited for this application. It limits surge current by functioning as a power resistor which drops from a high cold resistance to a low hot resistance when heated by the current flowing through it.

Some of the factors to consider when designing the NTC thermistor as an inrush current limiter are:

- Maximum permissible surge current at turn-on.
- Matching the thermistor to the size of the filter capacitors.
- Maximum value of steady state current.
- Maximum ambient temperature.
- Expected life of the power supply.

The main purpose of limiting inrush current is to prevent components in series with the input to the switcher from being damaged. Typically, inrush protection prevents nuisance blowing of fuses or breakers as well as welding of switch contacts. Since most thermistor materials are very nearly ohmic at any given temperature, the minimum no-load resistance of the thermistor is calculated by dividing the peak input voltage by the maximum permissible surge current in the power supply.

At the moment the circuit is energized, the filter capacitors in a switcher appear like a short circuit which, in a relatively short period of time, will store an amount of energy equal to $1/2CV^2$ through the thermistor. The net effect of this large current surge is to increase the temperature of the thermistor very rapidly during the period the capacitors are charging. The amount of energy generated in the thermistor during this capacitor charging period is dependent on the voltage waveform of the source charging the capacitors. However, a good approximation for the energy generated by the thermistor during this period is $1/2CV^2$ (energy stored in the filter capacitor). The ability of the NTC thermistor to handle this energy surge is almost entirely a function of the mass of the device.

During the short time that the capacitors are charging (usually less than 0.1 second), very little energy is dissipated. Most of the input energy then must be handled by the heat capacity (H) of the thermistor.

The maximum steady-state current rating of a thermistor is mainly determined by the acceptable life of the final products for which the thermistor becomes a component. As more current flows through the device, its steady-state operating temperature will increase and its resistance will decrease. So, rather than being a problem of maximum current, the problem becomes one of maximum temperature.

Chapter 9

PTC Thermistors

PTC (Positive Temperature Coefficient) thermistors are thermally sensitive resistors made of polycrystalline ceramic materials. The base compounds, usually barium titanate or solid solutions of barium and strontium titanate, are high-resistivity materials made semiconductive by the addition of dopants (impurities to alter properties of a pure substance).

Like most semiconductors, the PTC exhibits a slight negative temperature coefficient over the majority of its normal operating temperature region. Above the Curie temperature of the material, however, rapid changes in the ferro-electric properties of the ceramic causes a sharp rise in resistance - usually several orders of magnitude.

Electrical Characteristics

There are four electrical characteristics of the PTC thermistor:

- **Zero Power Resistance @ 25°C** - the measured resistance value of the PTC at 25°C with no appreciable self-heat producing current flowing through the thermistor. For practical purposes, most specifications call for a maximum measuring power of 0.1 mW.
- **Minimum Resistance** - defined as the lowest zero power resistance value that the thermistor is able to assume. It is the point on the

characteristic curve where the slope of the tangent line changes from negative to positive.

- **Transition Temperature** - the temperature at which the resistance of the PTC thermistor begins to rise sharply. It coincides roughly with the Curie point of the material. For specification purposes, some manufacturers define the transition temperature to be the temperature at which the resistance of the PTC is double the minimum resistance value.
- **Temperature Coefficient** - a measure of the slope of the PTC resistance/temperature curve at any point with respect to the resistance at that point. Below minimum resistance, the PTC thermistor exhibits a negative temperature coefficient. Above the transition temperature the positive temperature coefficient is a function of the material.

Thermal Characteristics

The PTC thermistor can operate in two modes: self-heated and nonself-heated. Self-heating is an operating condition where the temperature of the PTC is increased above the ambient by joule (I^2R) heating within the ceramic. The thermal characteristics of the PTC are described by the following terms:

- **Heat Capacity (H)** - is the amount of heat that is required to change the body temperature of a thermistor by one degree. PTCs have a heat capacity per unit volume of approximately 50 watt-sec./in^3/°C.
- **Dissipation Constant** - is the proportionality constant between wattage applied to a thermistor and its consequent temperature increase due to self-heating. Since the temperature rise of the PTC is a measure of the efficiency of the unit in shedding heat to its surroundings, and factor which affects heat flow will affect the dissipation constant. This includes lead material, mounting method, ambient temperature, presence of gas or liquid flow, and the geometry of the device itself.
- **Volt-Amp (V/I) Curve** - defines the relationship between current and voltage at any point of the thermal equilibrium. The temperature and the resistance of the PTC are affected by both power

dissipation (self-heating) and ambient conditions. Any factor that changes the dissipation constant also changes the shape of the volt-amp curve. One important point is the limit current point. It defines the amount of steady-state current the thermistor is able to pass before going into the high resistance (current limiting) mode.

• **Thermal Time Constant** - refers to the time required for the temperature of a self-heated PTC to change 63.2% of the difference between the self-heated temperature and the ambient temperature after the power is disconnected, usually in a no air-flow ambient.

Another consideration is voltage dependence. Voltage dependence is characteristic of PTC ceramic materials. For any PTC thermistor maintained at a constant temperature, resistance decreases as the voltage across the device increases. It is not uncommon for a resistance due to voltage dependence to decrease by more than one order of magnitude.

One example of the effects of voltage sensitivity on the PTC thermistor operation is the distortion created in the steady-state AC current waveform. The distortion occurs because the magnitude of the voltage across the PTC as it moves through the zero-crossing point region is much less than the peak voltage. This distorted sinusodial waveform can make it very difficult to measure rms current or power. An accurate measurement can be made with a true rms meter.

Operating voltage is the maximum normal steady-state voltage to guarantee the long-term stability and service life of the PTC thermistor. While the operating voltage ratings have a substantial safety margin built in, it is possible to cause PTC overvoltage failure by exceeding the positive temperature coefficient region and driving the PTC into a negative temperature zone. This condition will almost certainly cause a catastrophic failure. In most cases, it will take at least twice the rated voltage before this failure occurs. It is not recommended using the PTC at steady-state voltages greater than operating voltage plus 10%.

Application Information

The dramatic rise in resistance of a PTC at the transition temperature makes it an ideal candidate for current limiting applications. For currents below the limiting current, the power being generated in the unit is insufficient to heat the PTC to its transition temperature. However, when abnormally high-fault current flow, the resistance of the PTC increases at such a rapid rate that any further increase in power dissipation results in a reduction in current.

The PTC's ability to operate as a resettable fuse is very much a function of ambient conditions. To assure that the PTC will limit current under fault conditions and will not limit current under normal conditions, the following design parameters should be considered:

- The minimum amount of current required by the PTC to guarantee switching at the minimum ambient temperature.
- The maximum amount of current the PTC must be able to pass without switching at the maximum ambient temperature.
- The amount of time it takes the PTC to switch after subjected to a fault current.

Typical applications for PTC resettable fuses:

- Telephone line fault protection
- Fuse for intermittent solenoids
- Over-current protection for transformers
- Locked rotor protection for motors
- Transistor protection
- Speaker protection.

The PTC over-current protector is connected in series with the load which is to be protected. During normal operating conditions, the PTC remains in its low resistance state resulting in negligible attenuation to current flow. When a short circuit or over-current condition occurs, the PTC will switch into its high-resistance state thereby limiting the current flow in the circuit to a point well below the normal operating level. When the fault condition is removed, the PTC will return to its low resistance state allowing the current flow to recover to its normal level.

PTC thermistors provide an economical method of temperature control by combining the function of the heater and thermostat in one ceramic pellet. This thermostatic action is accomplished by the steep PTC characteristic. Voltage fluctuations are compensated for by a corresponding change in current. Even ambient temperature excursions are taken into account. For example, as ambient temperature rises, PTC resistance increases, thereby, reducing input power to the circuit.

Typical applications for PTC heaters:

- Contact lens cleaner kits
- Medical equipment
- In-line diesel fuel heater
- Temperature stabilization of electronic components
- LCD heaters.

Time delay, motor starting, and automatic picture tube degaussing are separate applications that are related to each other in this one respect: the PTC function is to allow an action to occur for a predetermined period of time, and then to halt that action until the circuit is de-energized.

Time delay circuits with the PTC in series with the relay coil function when the switch is closed, the relay will energize instantaneously and remain energized until the PTC switches to its high resistance state. When the PTC is in parallel with the relay coil, switch closed, the relay will not energize until the PTC switches to its high resistance state.

With a PTC in series with the starting winding in a single phase electric motor circuit, the PTC acts as a time delay device by replacing the trouble-some starting switch in a single-phase motor. When the circuit is energized, the PTC has a low resistance and permits most of the line voltage to be applied to the starter winding. As the motor starts, the PTC heats up until the transition temperature is reached at a rate determined by the thermal inertia of the PTC and current flowing through the starting winding. At this point, the thermistor rapidly changes from a low-resistance device to a high-resistance unit, and the steady-state power applied to the PTC is very low, causing the current through the starting winding to be negligible.

When the switch on the circuit with the PTC in series with a degaussing coil for a CRT in a color television or monitor is closed, a high percentage of the line voltage is impressed across the demagnetization coil. Within a short period of time, the PTC self-heats, switching from a low resistance level to a high resistance, and thereby decreasing the demagnetization coil current to near zero. The length of time for this switching action to occur is a function of the power applied to the PTC and its thermal inertia.

Air flow and liquid level sensing applications are based on the principle that the dissipation constant of a PTC changes with different ambient conditions. The net result of this changing dissipation constant is that greater heat losses either lower PTC temperatures or increase the power required to maintain a given temperature.

Unlike the NTC thermistor with its ability to sense temperature accurately over a long temperature span, the PTC thermistor is only useful as a measuring device at or near the transition temperature. Its extremely high temperature coefficient of resistance (in some cases as high as 200%/°C in the transition temperature region) allows the use of lower cost, less sophisticated electronics than is required with lower coefficient NTC thermistors.

In some applications, the PTC can be used as both the temperature sensor and the over-temperature protector. For example, the PTC can be utilized to control the charge cycle of rechargeable batteries. The low resistance of the PTC allows the battery to recharge normally. However, as the battery becomes fully charged, its temperature increases, at which point the PTC resistance increases dramatically to cut the charging current to a very low level.

In other applications, the PTC can be a triggering device by using the voltage across the thermistor as input to a comparator or amplifier. Surface mount PTC temperature sensors are excellent choices for these applications.

Surface mount PTC over-temperature protectors are designed to sense the temperature of power transistors, heat sinks, chassis, etc. Construction consists of a small (0.2") diameter PTC thermistor disc attached to an anodized aluminum mounting tab using thermally conductive epoxy. These devices remain in their low resistance state at all temperatures below the switch temperature. When the switch temperature is reached or exceeded, they increase in resistance rapidly thereby limiting current to the drive circuitry to protect the critical components. Once the temperature decreases to a normal operating level, the device resets to its low resistance state.

PTCs are generally designed to exhibit sharp increases in resistance at and above the switch temperature. However, special PTCs with nearly linear resistance-temperature characteristics are available for temperature compensation and temperature measurement applications.

The nearly constant temperature of a self-heated PTC results in a reduction in current through the unit as the voltage across the PTC is increased. By connecting the approximately constant-temperature PTC in parallel with a resistor, nearly constant load current can be obtained over a broad voltage range.

With a PTC placed across a switch, the PTC changes from a low resistance to a high resistance when the switch is opened. The low resistance of the PTC provides effective arc suppression, and the subsequent PTC switching action transfers essentially all of the power supply voltage from the inductive load to the PTC element.

Glossary

Ambient Operating Temperature - The temperature of the air surrounding an object, neglecting small localized variations.

Contact Resistance Variation - The apparent resistance seen between the wiper and resistance element when the wiper is energized with specified current and moved over the adjustment travel in either direction at a constant speed. The output variations are measured over a specified frequency bandwidth, exclusive of the effects due to roll-on or roll-off of the terminations and is expressed in ohms or percent of total nominal resistance.

Critical Value of Resistance - For a given voltage rating and a given power rating, there is only one value of resistance that will dissipate full rated power at rated voltage. This value of resistance is commonly referred to as the "critical value of resistance." For values of resistance below the critical value, the maximum (element) voltage is never reached and for values of resistance above critical value, the power dissipated becomes lower than rated.

Curie point - In ferroelectric dielectrics, the temperature at which the dielectric constant reaches peak values. At the Curie point, the crystal form is changing from cubic to tetagonal.

Current Noise - An AC component of voltage appearing across a resistor when current is passed through it. Usually expressed in RMS microvolts (μV) per volt applied to the resistor, it may also be expressed in noise index figures of dB.

Current Sensor - A resistive device employed to sense levels of current.

Current Shunt - A four-terminal shunt.

Dielectric Strength - The ultimate breakdown voltage of the dielectric or insulation of the resistor when the voltage is applied between the case and all terminals tied together. Dielectric strength is usually specified at sea level and simulated high altitude air pressures.

Dissipation Constant - The power required to raise the thermistor body by one degree Celsius. Expressed in mW/°C.

Heat Capacity (H) - The amount of heat that is required to change the body temperature of a thermistor by one degree Celsius.

Hot-spot Temperature - The maximum temperature measured on the resistor due to both internal heating and the ambient operating temperature. Maximum hot-spot temperature is predicated on thermal limits of the materials and the design. The hot-spot temperature is also usually established as the top temperature on the derating curve at which the resistor is derated to zero power.

Impedance - The ratio of effective voltage over effective current in an AC circuit.

Insulation Resistance - The DC resistance measured between all terminals connected together and the case, exterior insulation, or external hardware.

Maximum (element) Working Voltage ($E = PR$) - The maximum voltage stress (dc or rms) that may be applied to the resistor (resistance element) is a function of:

- the materials used
- the required performance
- the physical dimensions.

Minimum Resistance - The lowest zero-power resistance value that the PTC thermistor is able to assume.

Negative Temperature Coefficient (NTC) Thermistor - A semiconductor device that decreases in resistance as a result of an increase in body temperature.

Noise - An unwanted voltage fluctuation generated within the resistor. Total noise of a resistor always includes "Johnson" noise which is dependent only on the resistance value and temperature of the resistance element. Depending on the type of element and construction, total noise may also include noise caused by current flow, and noise caused by cracked bodies and loose end caps or leads. For variable resistors, noise may also be caused by jumping of contact over turns of wire and by an imperfect electrical path between the contact and resistance element.

Positive Temperature Coefficient (PTC) Thermistor - A device whose zero-power resistance increases with an increase in temperature.

Reactance - That component of an AC impedance defining a phase shift between the current and the voltage.

Resistance Ratio - The ratio of measured resistance at any two reference temperature points.

Resistance Temperature Characteristic (Temperature Coefficient) - The magnitude of change in resistance due to temperature usually expressed in percent per degree Celsius ($\%/°C$) or parts per million per degree Celsius ($ppm/°C$). If the changes are linear over the operating temperature range, the parameter is known as "temperature coefficient."

Resistance Tolerance - The permissible deviation of the manufactured resistance value (expressed in percent) from the specified nominal resistance value at standard (or stated) environmental conditions.

Resistor - A device that converts electrical energy to thermal energy according to Ohm's Law.

Shunt - A resistive device employed to divert most of the current in an electric circuit.

Stability - The overall ability of a resistor to maintain its initial resistance value over extended periods of time when subjected to any combination of environmental conditions and electrical stresses.

Temperature Coefficient - A measure of the slope of the resistance/temperature (R/T) curve at any point with respect to the resistance at that point. The temperature coefficient is commonly expressed as: %/°C, or ppm/°C.

Thermal Electromotive Force (Thermal EMF) - The temperature dependent voltage output that is produced continuously from two dissimilar metals or alloys joined together to form a metallurgical junction or thermocouple. Various metals and their alloys have different output levels in combination with other metals and alloys according to their position in the periodic table of elements. Since resistors commonly have copper leads, the various metals employed in making resistors have their thermal EMF's identified in μV/°C vs. copper.

Thermal Time Constant - The time required for the temperature of a self-heated PTC to change 63.2% of the difference between the self-heated temperature and the ambient temperature after the power is disconnected usually in a no airflow ambient.

Thermistor - contraction of "Thermally Sensitive Resistor" - a device that exhibits a predictable change in resistance as a function of any change in absolute body temperature.

Transition Temperature - The temperature at which the resistance of the PTC thermistor begins to rise sharply. It coincides roughly with the Curie point of the material.

Volt-Amp (*V/I*) Curve - The relationship between current and voltage at any point of the thermal equilibrium of a PTC.

Zero-Power Resistance - The resistance of a thermistor with zero electrical power dissipation or self-heating producing current flow through the thermistor.

Bibliography

Many thanks to the following who gave permission to use their information for reference, or for incorporation into this book:

Dale Electronics, Inc. (A subsidiary of Vishay Intertechnology, Inc.), Columbus, NE 68602: Full Line Catalog, Volume 7, Revision 919, A Comprehensive Guide to Resistive Shunts and Current Sensors.

Ketema-Rodan Division, Anaheim, CA 92806, Thermistor Product Guide, 1993 issue.

Keystone Carbon Co., St. Marys, PA 15857, NTC & PTC Thermistors & Probe Assemblies, 1992 issue.

Meedijk, V., "Selecting the Best Resistor/Capacitor." Radio Electronics, Gernsback Publications, Inc., Farmingdale, NY 11735, February, 1985, pp 47-50.

Sensor Scientific, Inc., Fairfield, NJ 07004, NTC Thermistors, Your Thermistor Planning Guide, 1993 issue.

Vishay Intertechnology, Inc., Malvern, PA 19355, Foil Resistors, Data Book 2000A.

Additional Reference:

United States Military Standard: MIL-STD-199C, Resistors, Selection and Use of, August 28, 1981.

Appendix A

*Resistor Selection Guidelines**

This is a summarization of resistor types available, their characteristics, recommended applications, and suggested derating factor. Use of a derating factor is an effective means to decrease the failure rate of most devices since life is stress and temperature dependent. Derating is accomplished by either decreasing part stresses such as power/voltage or current or by selecting a higher rated part. Optimum derating occurs at or below the point where an increase in stress or operating temperature results in a large increase in the device failure rate.

One note about the following information: The values and ratings shown are provided as guidelines. While they apply to the most commonly found units, it is not impossible to find units with slightly, or greatly, different specifications. Refer to manufacturer's data for specific ratings and specifications.

Carbon Composition

Resistance range: 2.7 ohms to 100 megohms
Power rating: to 2 watt
Tolerance: 5% to 20%
Temperature coefficient: -200 to -8000 ppm/°C.
Noise: less than 6 μV/V
Derating factors: 50% power, 80% voltage.
Notes: General purpose. Excellent transient and surge handling capabilities. RF produces capacitive effects end to end, and operation at VHF or higher frequencies reduces effective resistance due to dielectric losses. Resistance increases by 20% during storage under humid conditions.

---

* Reprinted with permission from **Radio-Electronics** Magazine, February 1985 issue.
© Copyright Gernsback Publications, Inc., 1985.

Carbon Composition Potentiometer

Resistance range: 50 ohms to 10 megohms
Power rating: to 5 watts
Temperature coefficient: +1000 ppm/°C.
Derating factors: 50% power, 80% voltage.
Life expectancy: 5,000,000 rotations
Failure mode: noise
Notes: High shaft torque causes poor adjustability.

Carbon Film

Resistance range: 10 ohms to 25 megohms
Power rating: 0.1 to 10 watts
Tolerance: 2% to 10%
Temperature coefficient: -200 to -1000 ppm/°C.
Noise: less than 10 μV/V. Derating factors: 50% power, 80% voltage.
Notes: General purpose, cost less than carbon-composition units.

Metal Film

Resistance range: 10 ohms to 3 megohms (high voltage types: 1 kilohm to 30 gigohms)
Power rating: to 10 watts (high voltage types: to 6 watts)
Tolerance: 0.1% to 20%
Temperature coefficient: ±25 to ±175 ppm/°C.
Noise: less than 0.1 μV/V
Life expectancy (potentiometers): 100,000 rotations
Failure mode: resistance change or catastrophic failure.
Derating factors: 50% power, 80% voltage.
Notes: Fair degree of precision in lower value units. High stability, long life, and excellent high-frequency performance. Resistance values stable to about 100 MHz; begin to decrease beyond that frequency. Used in high-frequency tuning circuits, measuring circuits, filters, etc.

Film Networks

Resistance range: 10 ohms to 33 megohms
Power rating: to 0.2 watts per element, to 1.6 watts per network.
Tolerance: 0.1% to 5%
Operating temperature range: -55 to +125°C.
Temperature coefficient: ±25 to ±300 ppm/°C.
Notes: Tracking between resistors 5 ppm/°C.

Chip Resistors

Resistance range: 1 ohm to 100 megohms
Power rating: to 2 watt
Tolerance: 1% to 20%
Operating temperature range: -55 to +125°C.

Power Wirewound

Resistance range: 0.1 ohm to 180 kilohms
Power rating: to greater than 225 watts
Tolerance: 5% to 10%
Temperature coefficient: less than ±260 ppm/°C.
Noise: low static, high dynamic noise levels.
Derating factors: 50% power, 80% voltage.

Precision Wirewound

Resistance range: 0.1 ohm to 800 kilohms
Power rating: to 15 watt
Tolerance: 0.01% to 1%
Temperature coefficient: varies with resistance
Noise: low static, high dynamic noise levels.
Life expectancy (potentiometers): 200,000 to 1,000,000 rotations
Failure mode: catastrophic failure
Derating factors: 50% power, 80% voltage.
Notes: Wirewound resistors are used in low-tolerance, high-power dissipation applications where AC performance is not critical. Power dissipation depends on the heat sink or air flow around the device. When mounting on a PC board, standoffs should be used to prevent charring the board. Not suitable for use a frequencies above 50 kHz. Wirewound potentiometers do not suffer from contact resistance variations. The units can be manufactured with low temperature coefficients and tight tolerances. Applications include motor speed controls, lamp dimmers, heater controls, etc. Precision types are used in servo mechanisms.

Hybrid Potentiometers

Resistance range: 200 ohms to 250,000 ohms
Power rating: to 7 watts
Tolerance: 5%
Temperature coefficient: less than ±100 ppm/°C.
Life Expectancy: 10,000,000 rotations

Cermet

Resistance range: 50 ohms to 5 megohms
Power rating: to 2 watts
Life expectancy (potentiometers): 50 to 500,000 rotations
Failure mode: noise
Derating factors: 50% power, 80% voltage.
Notes: very stable under humid conditions. Low temperature coefficients. Low end resistance (2 ohms). Short life expectancy. High resolution of the resistive element allows for more precise trimmer settings. Less reactance in high-frequency applications than wirewound units, and are lower in price. Cermet is also the thick film used in resistor networks and chip resistors.

Conductive Plastic Potentiometers

Resistance range: 150 ohms to 5 megohms
Power rating: to 1 watt
Temperature coefficient: -600 to -300 ppm/°C.
Life expectancy: 100,000 to 4,000,000 rotations
Failure mode: noise
Derating factors: 50% power, 80% voltage.

Precision Conductive Plastic Potentiometers

Resistance range: 100 ohms to 500 kilohms
Power rating: to 7 watts
Tolerance: 3%
Temperature coefficient: less than 70 ppm/°C.
Life expectancy: Greater than 2,000,000 rotations.

Conductive Plastic Trimmers

Resistance range: 10 ohms to 100,000 ohms
Power rating: to 1 watt
Notes: Conductive plastic potentiometers have a long life expectancy and low-noise characteristics. Resistance will shift if exposed to humidity.

Foil

Notes: Foil resistors can be used to replace metal film, wirewounds, trimmers, and potentiometers, where ultra precision, long life, and circuit stability in extreme environmental conditions are required.

Appendix B

Symbols and Equations

Symbols

E = Voltage in *Volts*

I = Current in *Amperes*

R = Resistance in *Ohms*

P = Power in *Watts*

C = Capacitance in *Farads*

L = Inductance in *Henries*

X_C = Capacitive Reactance in *Ohms*

X_L = Inductive Reactance in *Ohms*

Metric Prefixes

Pico	$\times 10^{-12}$	Tera	$\times 10^{12}$
Nano	$\times 10^{-9}$	Giga	$\times 10^{9}$
Micro	$\times 10^{-6}$	Mega	$\times 10^{6}$
Milli	$\times 10^{-3}$	Kilo	$\times 10^{3}$
Deci	$\times 10^{-1}$	Deca	$\times 10^{1}$

Basic Resistor Formulas

1. Ohm's Law

$$E = IR = \frac{P}{I} = \sqrt{PR} \qquad I = \frac{E}{R} = \frac{P}{E} = \sqrt{\frac{P}{R}}$$

$$R = \frac{E}{I} = \frac{E^2}{P} = \frac{P}{I^2} \qquad P = I^2 R = \frac{E^2}{R} = EI$$

2. Resistance in Series

$$R_T = R_1 + R_2 + R_3 + R_4 + \ldots$$

3. Resistance in Parallel

$$\frac{1}{R_T} = \frac{1}{R_1} + \frac{1}{R_2} + \frac{1}{R_3} + ...$$

$$R_T = \frac{R_1 \times R_2}{R_1 + R_2} \quad (For\, 2\, Resistors\, Only)$$

$$R_T = \frac{R_1}{\#\, of\, Resistors} \quad (For\, Equal\, Resistors)$$

4. Capacitive Reactance

$$X_C = \frac{1}{2\,\pi\,f\,C} \quad (f = Frequency\, in\, Hz)$$

5. Inductive Reactance

$$X_L = 2\,\pi\,f\,L \quad (f = Frequency\, in\, Hz)$$

6. Thermal Noise

$$E^2 = 4KTR(f2 - f1)$$

where:

K = BOLTZMANN'S constant (1.38 x 10^{-23} Joules/Degree Kelvin)
T = Absolute Temperature ($^{\circ}$C + 273)
R = Resistance of the conductor
E = Root mean square value of the thermal noise voltage
(f2 -f1) = Bandwidth in Hz

7. Noise Index

$$db = 20 \times \log_{10} \frac{Noise\, Voltage\, (over\, a\, 1\, decade\, bandwith)}{DC\, Voltage}$$

8. Parts Per Million (ppm): Conversion of % to ppm

%	ppm	%	ppm
0.0001%	1	0.01%	100
0.0002%	2	0.02%	200
0.0005%	5	0.025%	250
0.001%	10	0.05%	500
0.0025%	25	0.1%	1,000
0.005%	50	1.0%	10,000

Formula is: $\dfrac{0.0001\% \times 10^6}{100} = 1\, ppm$

Resistor Color Code

0 = Black	6 = Blue
1 = Brown	7 = Violet
2 = Red	8 = Grey
3 = Orange	9 = White
4 = Yellow	5% = Gold (or $x10^{-1}$)
5 = Green	10% = Silver (or $x10^{-2}$)
	20% = No color

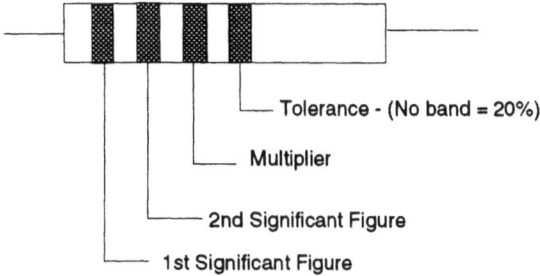

Tolerance - (No band = 20%)

Multiplier

2nd Significant Figure

1st Significant Figure

Commercial 4 Band

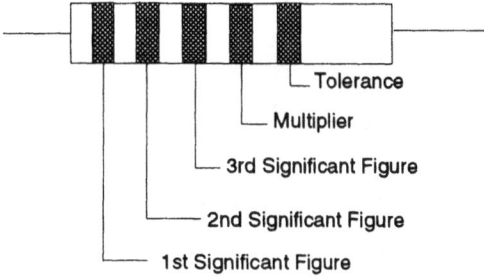

Tolerance

Multiplier

3rd Significant Figure

2nd Significant Figure

1st Significant Figure

Commercial 5 Band - 1%

Standard Resistance Values

0.1% 0.25% 0.5%	1%	0.1% 0.25% 0.5%	1%	0.1% 0.25% 0.5%	1%	0.1% 0.25% 0.5%	1%
10.0	10.0	20.0	20.0	40.2	40.2	80.6	80.6
10.1		20.3		40.7		81.6	
10.2	10.2	20.5	20.5	41.2	41.2	82.5	82.5
10.4		20.8		41.7		83.5	
10.5	10.5	21.0	21.0	42.2	42.2	84.5	84.5
10.6		21.3		42.7		85.6	
10.7	10.7	21.5	21.5	43.2	43.2	86.6	86.6
10.9		21.8		43.7		87.6	
11.0	11.0	22.1	22.1	44.2	44.2	88.7	88.7
11.1		22.3		44.8		89.8	
11.3	11.3	22.6	22.6	45.3	45.3	90.9	90.9
11.4		22.9		45.9		92.0	
11.5	11.5	23.2	23.2	46.4	46.4	93.1	93.1
11.7		23.4		47.0		94.2	
11.8	11.8	23.7	23.7	47.5	47.5	95.3	95.3
12.0		24.0		48.1		96.5	
12.1	12.1	24.3	24.3	48.7	48.7	97.6	97.6
12.3		24.6		49.3		98.8	
12.4	12.4	24.9	24.9	49.9	49.9		
12.6		25.2		50.5			
12.7	12.7	25.5	25.5	51.1	51.1		
12.9		25.8		51.7			
13.0	13.0	26.1	26.1	52.3	52.3		
13.2		26.4		53.0			
13.3	13.3	26.7	26.7	53.6	53.6		
13.5		27.1		54.2			
13.7	13.7	27.4	27.4	54.9	54.9		
13.8		27.7		55.6			
14.0	14.0	28.0	28.0	56.2	56.2		
14.2		28.4		56.9			
14.3	14.3	28.7	28.7	57.6	57.6		
14.5		29.1		58.3			
14.7	14.7	29.4	29.4	59.0	59.0		
14.9		29.8		59.7			
15.0	15.0	30.1	30.1	60.4	60.4		
15.2		30.5		61.2			
15.4	15.4	30.9	30.9	61.9	61.9		
15.6		31.2		62.6			
15.8	15.8	31.6	31.6	63.4	63.4		
16.0		32.0		64.2			
16.2	16.2	32.4	32.4	64.9	64.9		
16.4		32.8		65.7			
16.5	16.5	33.2	33.2	66.5	66.5		
16.7		33.6		67.3			
16.9	16.9	34.0	34.0	68.1	68.1		
17.2		34.4		69.0			
17.4	17.4	34.8	34.8	69.8	69.8		
17.6		35.2		70.6			
17.8	17.8	35.7	35.7	71.5	71.5		
18.0		36.1		72.3			
18.2	18.2	36.5	36.5	73.2	73.2		
18.4		37.0		74.1			
18.7	18.7	37.4	37.4	75.0	75.0		
18.9		37.9		75.9			
19.1	19.1	38.3	38.3	76.8	76.8		
19.3		38.8		77.7			
19.6	19.6	39.2	39.2	78.7	78.7		
19.8		39.7		79.6			

2% 5%	10%
10	10
11	
12	12
13	
15	15
16	
18	18
20	
22	22
24	
27	27
30	
33	33
36	
39	39
43	
47	47
51	
56	56
62	
68	68
75	
82	82
91	

Standard resistance values are obtained from the decade table by multiplying by powers of 10. As an example, 13.3 can represent ohms, 133 ohms, 1.33K, 13.3K, 133K, or 1.33 megohms.

Index

Notes

Notes